Translational Systems Sciences

Volume 3

In 1956, Kenneth Boulding explained the concept of General Systems Theory as a *skeleton of science*. He describes that it hopes to develop something like a "spectrum" of theories—a system of systems which may perform the function of a "gestalt" in theoretical construction. Such "gestalts" in special fields have been of great value in directing research towards the gaps which they reveal.

There were, at that time, other important conceptual frameworks and theories, such as cybernetics. Additional theories and applications developed later, including synergetics, cognitive science, complex adaptive systems, and many others. Some focused on principles within specific domains of knowledge and others crossed areas of knowledge and practice, along the spectrum described by Boulding.

Also in 1956, the Society for General Systems Research (now the International Society for the Systems Sciences) was founded. One of the concerns of the founders, even then, was the state of the human condition, and what science could do about it.

The present Translational Systems Sciences book series aims at cultivating a new frontier of systems sciences for contributing to the need for practical applications that benefit people.

The concept of translational research originally comes from medical science for enhancing human health and well-being. Translational medical research is often labeled as "Bench to Bedside." It places emphasis on translating the findings in basic research (*at bench*) more quickly and efficiently into medical practice (*at bedside*). At the same time, needs and demands from practice drive the development of new and innovative ideas and concepts. In this tightly coupled process it is essential to remove barriers to multi-disciplinary collaboration.

The present series attempts to bridge and integrate basic research founded in systems concepts, logic, theories and models with systems practices and methodologies, into a process of systems research. Since both bench and bedside involve diverse stakeholder groups, including researchers, practitioners and users, translational systems science works to create common platforms for language to activate the "bench to bedside" cycle.

In order to create a resilient and sustainable society in the twenty-first century, we unquestionably need open social innovation through which we create new social values, and realize them in society by connecting diverse ideas and developing new solutions. We assume three types of social values, namely: (1) values relevant to social infrastructure such as safety, security, and amenity; (2) values created by innovation in business, economics, and management practices; and, (3) values necessary for community sustainability brought about by conflict resolution and consensus building.

The series will first approach these social values from a systems science perspective by drawing on a range of disciplines in trans-disciplinary and cross-cultural ways. They may include social systems theory, sociology, business administration, management information science, organization science, computational mathematical organization theory, economics, evolutionary economics, international political science, jurisprudence, policy science, socio-information studies, cognitive science, artificial intelligence, complex adaptive systems theory, philosophy of science, and other related disciplines. In addition, this series will promote translational systems science as a means of scientific research that facilitates the translation of findings from basic science to practical applications, and vice versa.

We believe that this book series should advance a new frontier in systems sciences by presenting theoretical and conceptual frameworks, as well as theories for design and application, for twenty-first-century socioeconomic systems in a translational and trans-disciplinary context.

More information about this series at http://www.springer.com/series/11213

Kuntoro Mangkusubroto • Utomo Sarjono Putro •
Santi Novani • Kyoichi Kijima
Editors

Systems Science for Complex Policy Making

A Study of Indonesia

Springer

Editors
Kuntoro Mangkusubroto
School of Business and Management
Institut Teknologi Bandung
West Java, Indonesia

Utomo Sarjono Putro
School of Business and Management
Institut Teknologi Bandung
West Java, Indonesia

Santi Novani
School of Business and Management
Institut Teknologi Bandung
West Java, Indonesia

Kyoichi Kijima
Tokyo Institute of Technology
Meguro-ku, Tokyo, Japan

ISSN 2197-8832   ISSN 2197-8840   (electronic)
Translational Systems Sciences
ISBN 978-4-431-56649-6   ISBN 978-4-431-55273-4   (eBook)
DOI 10.1007/978-4-431-55273-4

Printed on acid-free paper

This Springer imprint is published by Springer Nature
The registered company is Springer Japan KK

About This Book

This volume has applied a systems science perspective to complex policy-making dynamics, using the case of Indonesia to illustrate the concepts. Indonesia is an archipelago with a high heterogeneity. Its people consist of 1340 tribes that scattered over 17,508 islands. Every region has different natural strengths and conditions. In the national development process, all regions depend on one another while optimizing their own conditions. In addition to this diversity, Indonesia also employs a democratic system of government with high regional autonomy. A democratic government puts a high value on individual freedom, but on the other hand, it has caused conflicts of interest to occur more frequently. High regional autonomy also often causes problems in coordination among agencies and regional governments. This uniqueness creates a kind of complexity that is rarely found in other countries. These daily complexities require intensive interaction, negotiation processes, and coordination. Such necessities should be considered in public policy making and in managing the implementation of national development programs. In this context, common theories and best practices that generated on the basis of more simplified assumptions are often failed. Systems science offers a way of thinking that can take these issues into account and potentially overcome these complexities. However, the efforts to apply systems science massively and continuously in real policy making by involving many stakeholders are still rarely carried out. The first part of the book discusses the gap between the existing public policy-making approach and the needs in the real world. After that, the characteristics of the appropriate policy-making process in a complex environment and how this process can be carried are described. In later sections, important systems science concepts that can be applied in managing these complexities are discussed. Last but not least, the efforts to apply these concepts in real cases in Indonesia are described.

Contents

Supporting Decision Making for a Republic Under a Complex System

Kuntoro Mangkusubroto, Dhanan Sarwo Utomo, and Dyah Ramadhani

Abstract This article discusses the complexity of policy development at the national level in the real world. In this discussion, an action research conducted by the authors in the REDD+ case in Indonesia is presented. In this action research, we give descriptions on how to orchestrate the interaction among various actors at international, national and local level. These efforts have produced many benefits, e.g. various sustainable development programmes that have been implemented in various regions in Indonesia.

Introduction

Indonesia stretches from Aceh to Papua (one-eighth of the globe circle), consisting of more than 13.400 islands and territorial waters extending up to 3.257.483 km^2.[1] The archipelago contains more than 400 ethnics who speak different languages. The country's population has reached 245 million in 2014. Economically, the country has been developing briskly. The GDP was USD 870 billion in 2013, making it one of the 20 largest economies around the world. The GDP per capita (in constant 2005 USD) rose significantly from only USD 840 in the early 1990s to USD 1810 in 2013.[2] Indonesia may well be considered a giant compared to its neighbouring countries in Southeast Asia.

---

[1] Data from Badan Informasi Geospasial, Republik Indonesia. Accessed from http://tanahair.indonesia.go.id/home/ on 9th April 2015

[2] GDP is in constant 2005 US dollars. Source: World Development Indicators database, World Bank. Accessed from http://data.worldbank.org/indicator on 9th April 2015

K. Mangkusubroto (✉) • D.S. Utomo
School of Business and Management, Institut Teknologi Bandung, West Java, Indonesia
e-mail: m.kuntoro@gmail.com; dhanan@sbm-itb.ac.id

D. Ramadhani
Harvard Kennedy School, Cambridge, MA, USA
e-mail: dyah.ramadhani@gmail.com

© Springer Japan 2016
K. Mangkusubroto et al. (eds.), *Systems Science for Complex Policy Making*,
Translational Systems Sciences 3, DOI 10.1007/978-4-431-55273-4_1

1

Yet the immense growth does not translate into any distinction on how the government of Indonesia works compared to its peers. Government bureaucracy approach is sectoral in nature. Ministries are designed to fulfil only a single sector or purpose such as the Ministry of Infrastructure, Ministry of Health or Ministry of Energy. In reality, challenges are becoming more cross-sectoral and multidimensional. One cannot separate the issue of health from poverty nor solve an environmental issue without hinging on energy and education. Simply creating solutions for one sole purpose will deem to beat its own goal. The Indonesian government like most governments is still struggling to adapt to the complexity of problems where factors are intertwined more than ever.

As the fourth largest democracy, Indonesia is faced with definite coalition in the government and constant compromise in the cabinet. It is even more severe in a divergent political condition reflected by the multiparty system adopted by Indonesia. With fifteen (15) parties recognised in the 2014 election, the multiparty system requires the need to develop a coalition in the government, which led to compromise in day-to-day cabinet decision-making. One ruling government is usually consisted of four to five party coalitions with one to two major parties as opposition. Despite the 2009–2014 administration that has managed to have six supporting parties within the cabinet, the government coalition is only supported by four parties. Moreover, the winning party of 2014 controls only less than 30 % of the parliament. Thus, more compromise in both the executive body and legislative body is given. Following the reform in 1998, Indonesia has chosen to embark on democratisation and decentralisation at the same time. This is a momentous change in the country's political reform from a centralised and top-down system to a decentralised and bottom-up mechanism.

The people directly elect the president, governors, and heads of districts. Indonesia, now comprising 34 provinces and 506 municipalities and districts, conducts more elections for each of the governance levels from national to district leaders including their legislative counterparts.

This situation is complicated by the implementation of the decentralisation system whereby the local government has been given certain power and authority and the central government has no direct control over it. Starting with Regulation No. 22 of 1999 and subsequently with Regulation No. 32 of 2004, the central government's authority has shrunk into foreign affairs; defence; security; judicial, religious, and national monetary; and fiscal affairs. The sectoral approach is also mimicked by local governments as government budget and disbursement is structured and based on the existing ministries. Silo mindset and work process coupled with lack of understanding to intricacy of challenges led to poor delivery of public services and slow debottlenecking of obstacles, if any. Indonesia has potentially delegated more authority to local leaders than what they are now capable to manage.

One observed conclusion is for a policy to be made, whether in the local or national government; it needs more than just data and analysis. Policy is a product of rigorous interplay between actors, subject to their compromise and coalition.

Policy Making in National Complex

This section discusses the definition of the policy development process that is used in this chapter, its scope, and the stages in this process. The terms strategy and policy are often used interchangeably. In the strategic management literatures, for example, Rao et al. (2004) have stated that a strategy is defined as a comprehensive plan of action that is designed to meet specific objectives, in a certain time limit. On the other hand, a policy is defined as guidance for an organisation to make appropriate decisions in the long run. Although both terms imply a decision-making process, the term policy is often considered to have a broader meaning and also used in more fields. For simplicity, this paper uses the term policy in referring to the decision-making activities that are carried out to achieve the desired objectives.

At the national level, authors observed that there are at least three types of triggers for the policy development process, i.e.:

1. The need to improve the administration performance. For example, to improve the performance of the bureaucracy in Indonesia, the President established a working unit known as the President's Delivery Unit for Development Monitoring and Oversight (UKP4). The objectives of this unit are to oversee the progress of ministry and agency programmes, to synergise the working process across ministries and to solve the occurring bottleneck. The authors are actively involved in this unit.
2. The need to respond to state emergencies, for example, the reconstruction process in Indonesia (Aceh and Nias region) after the 2004 tsunami. Some of the authors were directly involved in the Rehabilitation and Reconstruction Agency for Aceh and Nias. This agency was responsible for managing international aid in the reconstruction effort. Some policies, such as the geographic information system to show how the aim is used, are made in an effort to transparently manage the received aid.
3. The need to respond to global challenges, for example, the policies that are made as efforts to address the climate change. Some of the authors are actively involved in the REDD+ task force which is responsible for designing and implementing policies of sustainable forest use.

Although the policy development process may have different triggers, the activities that are carried out and the requirements in each activity are generally the same. Summarising the information from several literatures, the policy development process can be depicted as a cycle as shown in Fig. 1.

Setting Direction This stage aims to decide the future that will be pursued, as well as issues and problems that can arise in achieving this desired future. In the public policy field, this activity is often known as agenda setting (Jann and Wegrich 2007). There are many reasons that may trigger this activity. In the public sector, this activity could be triggered when there are stakeholders who feel that the current

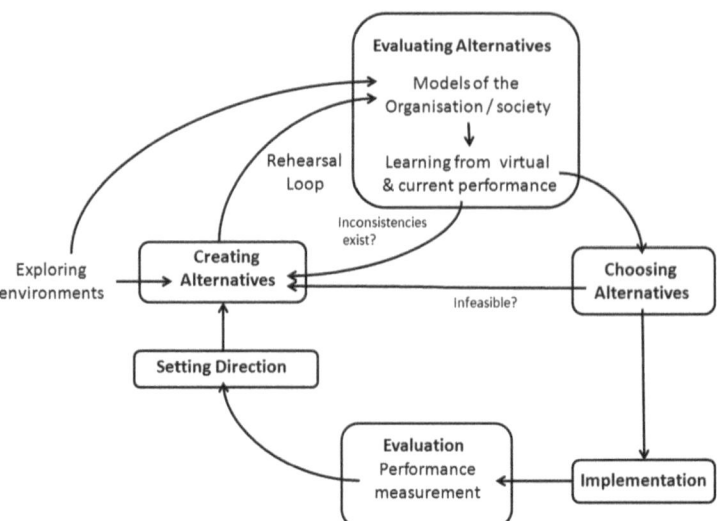

Fig. 1 Stages in policy development process

conditions do not meet their expectations or if there are several stakeholders whose interests are conflicting with each other (Jann and Wegrich 2007). In an organisation, this activity might be triggered when there are gaps between their vision, mission or goals and their performance (Dyson et al. 2007). These two reasons imply that the stakeholders already know the desired future or the conditions that may happen in the future. However, this activity may also be performed by stakeholders to design the future conditions that may be good and more desirable to them. The latter pattern is also known as interactive planning (Ackoff 1974; Ackoff 2001).

Creating Alternatives   This stage aims to identify various alternatives that may help in achieving the desired future. In public policy, this stage is often called as the policy formulation stage (Jann and Wegrich 2007). At this stage the stakeholders require information that may help them to understand the features and the uncertainties of the problem they are facing. This information gathering effort is often is known as shaping (Friend and Hickling 2005) or exploring internal and external environments (Dyson et al. 2007).

Evaluating Alternatives   In their book Friend and Hickling (2005) call this stage a comparing stage. In this stage, stakeholders construct and agree on the criteria that will be used (Friend and Hickling 2005) and then compare the performance of each alternative based on these criteria. This performance can be evaluated based on the social, environmental or economic impacts (Milano et al. 2014) of each of the alternatives, costs and benefits, the necessary coordination and conflict resolution if there are any (Jann and Wegrich 2007). This performance can also be measured by the potential increase in performance measurement of an organisation (Dyson et al. 2007). In evaluating available alternatives, stakeholders must be cautious because a complex policy problem often shows unexpected reactions to given interventions. Such reactions are often known as unintended consequences (Dyson et al. 2007) or policy resistance (Sterman 2001). Furthermore, an intervention often has irreversible effects after being applied in the real world and makes policy experimentation in such context not feasible. To avoid this, Dyson et al. (2007) suggested that the evaluation stage involves a rehearsal loop. In the rehearsal loop, the impacts of each alternative are projected using relevant models and information from the environment. The results of this projection may stimulate a virtual learning process that is useful to identify the inconsistencies and improve the existing alternatives (Dyson et al. 2007).

Choosing Alternatives   At this stage, the stakeholders agree upon the alternatives that will be performed as well as their priorities, to resolve at least some parts of the problem (Friend and Hickling 2005). There is also the possibility that some alternatives are left open, waiting for further developments in the situation. According to Pidd (2004), stakeholders actually realise that whatever they decide at this time may need to be revised in the future. In addition, the selected alternative is not necessarily the one with the best performance, but it is often a result of negotiations among diverse stakeholders (Jann and Wegrich 2007). Although it is often difficult to be achieved, the consensus among stakeholders is very important so that the selected alternative can be implemented. These reasons may require stakeholders to review and revise the available alternatives (Friend and Hickling 2005).

Implementation   At this stage, intervention is applied in the real world. In this process various changes in the management process within an organisation are carried out (Dyson et al. 2007). However, if the context involves many organisations, the coordination and interaction between intra- and inter-organisation also needs to be considered and managed (Jann and Wegrich 2007), in order to support a successful policy implementation.

Evaluation   The last stage is to evaluate whether the interventions can solve or reduce current problems (Jann and Wegrich 2007). This process can be carried out by measuring the improvement based on the existing performance measurement system (Dyson et al. 2007). The results will serve as feedback for future policy development processes (Dyson et al. 2007; Jann and Wegrich 2007).

Involving various departments, interest groups and society members is very important in this process, although the degree of their participation may vary based on the context (Jann and Wegrich 2007). Involving relevant external stakeholders is also considered to be able to improve the effectiveness of this process even in policy development for a single organisation (Dyson et al. 2007).

Pidd (2004) mentioned that there are two types of human rationality that play a role during a policy development process, namely, substantive rationality and procedural rationality. Simon (1986) described the substantive rationality as a decision-maker's effort to choose behaviours (alternatives) that are suitable to achieve a certain goal. In substantive rationality, the objectives and all possible alternatives are assumed to be known; the problem is how to select the best alternative. In contrast, the problem in procedural rationality is how to process limited information so that the decision-maker can make a choice (Simon 1986; Jones 2002). Much disjointed and episodic interaction in the policy development process could not be adequately explained without considering a behavioural model of human choice (Jones 2002).

Pidd (2004) suggested that in a policy development process, procedural rationality has a more dominant role when the stakeholders try to define their vision and to find out what can be done to achieve this vision. In this effort, the stakeholders need to make their worldview and values explicit and filter available information from the environment. In policy development stages discussed previously, these efforts were performed on the setting direction stage and creating alternatives stage. For example, the setting direction stage can become an enigma if the stakeholders do not appreciate what each other's roles, selective attention (stakeholder's ability to filter information) and emotional arousal play (Jones 2002).

Though its role is less dominant, the substantive rationality also contributes in suggesting how the data from the real world can be interpreted. After the possible alternatives are identified, substantive rationality is more suitable to be used, because it can objectively compare these alternatives based on the agreed criteria (Pidd 2004). These efforts are performed in evaluating and choosing alternatives stages that have been discussed previously. However, it also has been discussed previously that in the choosing alternatives stage the consensus among stakeholders needs to be established. Building this consensus involves negotiation that also requires the stakeholders to express their values and worldview. Therefore, procedural rationality can also play its role again at this stage.

According to Pidd (2004), approaches that may help the stakeholders in conducting information gathering, expressing their values and priorities and facilitating debate are required to support procedural rationality. These kinds of approaches enable the stakeholders to identify alternatives that can resolve the conflict. Meanwhile, to support the substantive rationality, approaches that can rank the alternatives objectively based on all the available information are required. This characteristic is usually owned by quantitative approaches.

Figure 2 describes how each type of rationality contributes in a policy development process.

In reality, it is very difficult to implement the structured process described in this section. The next section discusses the authors' efforts to carry out the process of policy development in a national context.

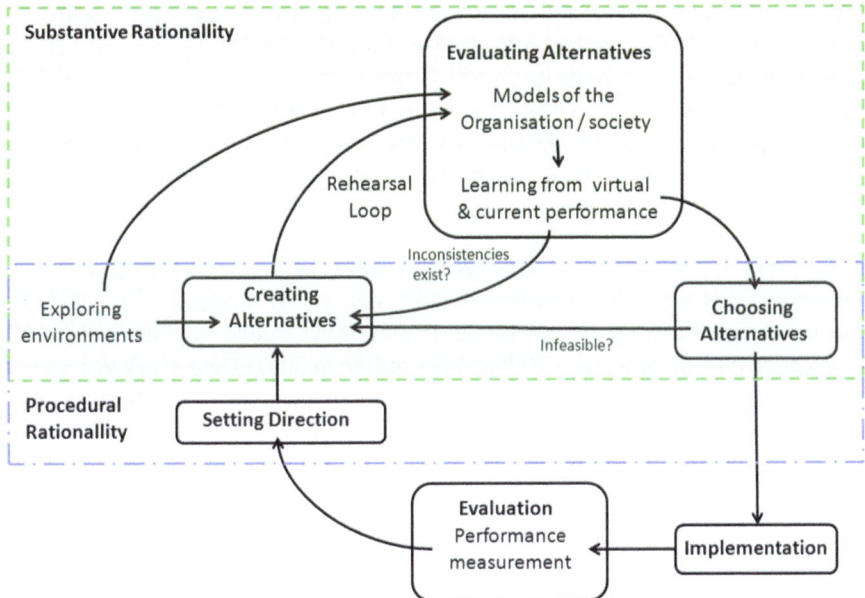

Fig. 2 Types of rationality mainly used by the stakeholders at each stage in policy development process

REDD+ Case in Indonesia

This section discusses the REDD+ case to describe the complexity of policy development in order to support the international agenda to tackle climate change. The first part discusses the underlying political conditions of REDD+ implementation in Indonesia. The second part discusses the actors involved in REDD+ implementation. The remaining parts discuss the efforts that have been made to orchestrate the interaction between actors at the international level, national level and local level.

REDD+ Politics in Indonesia

REDD+ was formally launched in Indonesia since the signing of the letter of intent (LoI) between the Indonesian government and Norwegian government in Oslo in May 2010. In this LoI, Norway hopes that the REDD+ fund should be managed credibly and transparently as was done by the Rehabilitation and Reconstruction Agency for Aceh and Nias (BRR) after the 2004 tsunami.

The LoI consists of three phases with USD 1 billion of total funding from Norway until 2020. The first phase is the preparation phase in which Indonesia should establish institutions that will manage REDD+. The second phase is the transformation phase wherein Indonesia should institutionally and operationally be ready to run a performance-based payment mechanism. The third phase is the phase in which REDD+ in Indonesia is run with performance-based payments, using USD 800 million from Norway. The terms in this LoI are a breakthrough in interstate partnership, whereby the developed countries no longer give 'aid' to the developing countries, but implement equal partnership to contribute to the emission reduction efforts in the developing countries.

The REDD+ itself is a relatively new concept. This concept began in the 1990s by providing incentives for the developing countries to protect their forests; hence, the carbon emissions from the forests can be reduced. This concept was officially adopted at the Conference of the Parties (COP) of the United Nations Framework Convention on Climate Change (UNFCCC) in Bali in 2007. Then, Norway became the initiator of this concept. Indonesia became a living laboratory for REDD+ since 2010 until today. The REDD+ concepts are continuously changed and altered due to experimentation made by Indonesia. From these experiences, Indonesia then introduces REDD+ as an instrument of development, not only to reduce emissions but also to improve welfare of local communities (beyond carbon and more than forest). The remaining discussion describes the actors and factors that play a role in REDD+.

The Actors Behind REDD+

The Norwegian Government

The Norwegian government is the pivotal actor that promotes REDD+ to the global stage and to Indonesia. At the 2007 COP in Bali, they launched a global initiative for forests and climate change and promised to contribute USD 500 million per year to the REDD+ initiative (this initiative is known as Norway's International Climate and Forest Initiative (NICFI)). With these efforts, they hope that they can play an important role internationally. Hence, they urged the REDD+ to become the key element of the COP in Bali.

The Indonesian Government

During the G-20 Summit in September 2009, the Indonesian government launched the ambitious emission reduction commitments. The Indonesian government mentioned that in 2020, Indonesia will be able to reduce emissions by 26% against the business-as-usual scenario by its own power and 41% with international support. The Norwegian government then stepped in and offered collaboration under the

REDD+ initiatives. Norway became the first country that signs a formal agreement with Indonesia.

The President's Delivery Unit for Development Monitoring and Oversight (UKP4) is then asked to coordinate with the REDD+ implementation in Indonesia. Some officials at this institution have been involved in the Aceh and Nias reconstruction after the tsunami in 2004. This institution also has the authority to coordinate with the ministries and agencies related to REDD+. Hence, the UKP4 is expected to be able to help in initiating the REDD+ implementation efforts in Indonesia.

The REDD+ agency (BP REDD+) is finally established in August 2013, after almost 3 years of preparations. The preparations of the presidential decree on the BP REDD+ itself require 1 year of discussion by involving various ministries and agencies. The operationalisation of the BP REDD+ itself is in fact very difficult. The head of BP REDD+ was chosen in December 2013, and its deputies are appointed in May 2014. This has suggested that developing a consensus to establish BP REDD+ was very difficult.

The International Interactions in REDD+ Dynamics

The Role of the United Nations

The UN's role in encouraging the implementation of REDD+ in Indonesia is very pivotal. This effort began with the UNFCCC, wherein the REDD+ was accepted as a key element in the 2007 COP. The United Nations then launched the UN REDD programme in 2008 by involving 56 partner countries. This programme was supported by three main UN agencies, e.g. the United Nations Development Programme (UNDP), Food and Agriculture Organization (FAO), and United Nations Environment Programme (UNEP). In this programme Indonesia becomes one of the partner countries. One of the pilot projects was carried out in the Central Sulawesi province by cooperating with the Indonesian Ministry of Forestry. The concepts adopted by the UN REDD programme are similar to the UNFCCC concepts for REDD+, i.e. giving incentives for the developing countries to reduce deforestation and forest degradation, as well as conservation and sustainable forest management, and to improve carbon stocks.

Indonesia has later managed to convince the UN that an effective coordination mechanism to implement REDD+ is necessary. As a result, the UN Secretary-General himself came to Palangkaraya and established the UN Office for REDD+ Coordination in Indonesia (UNORCID) in 2011. The UNORCID becomes the counterpart of the REDD+ task force and later the BP REDD+ in developing Indonesia's REDD+ agenda. The UNORCID is able to communicate the importance of Indonesia's REDD+ to the UN, and therefore, the United Nations has full support for Indonesia's REDD+ initiatives. The UNORCID has assisted Indonesia to convey the conception of REDD+ as 'beyond carbon and more than forest' in various world events. In various international conversations, the world began to see REDD+ not only as a climate change agenda but also as a sustainable development agenda.

The UNFCCC is the official forum for the global REDD+ since the 13th COP in Bali in 2007. However, the negotiations in this forum somehow do not represent interests of the developing countries, and this has made the REDD+ agenda progress very slowly. The deforestation and land use problems are more than just climate change issues. The main questions to these problems are how to manage forests and lands (governance issue) and how it may benefit the people (welfare issue). In addition to the climate change, the governance and welfare dimensions are very important for the developing countries including Indonesia. These two issues are often not considered in the UNFCCC.

At that time UNFCCC still cannot adopt Indonesia's innovations in REDD+. However, the collaboration between BP REDD+, Norway and UNORCID/UNDP was eventually able to disseminate REDD+'s innovation globally, e.g. the World Bank, Germany and the UK. Germany, for example, is interested in talking about REDD+ in the jurisdiction context. They created the REDD Early Movers programme. This is another innovation that involves development aspects in REDD+.

Besides these countries, there are many other international NGOs. For example, the Center for International Forestry Research (CIFOR), World Agroforestry Centre (ICRAF), World Resources Institute (WRI), Climate Advisers, etc. are research institutions who disseminate the REDD+ agenda globally. In the UN Climate Summit in September 2014, REDD+ managed to become a main issue in forest agenda discussion. This shows the importance of forest and climate agenda in the global development discourse.

One important milestone in the integration of REDD+ into the global development agenda is the involvement of the President of Indonesia as a co-chair at the High-Level Panel of Eminent Persons (HLPEP) on the Post-2015 Development Agenda. It's a long story, starting from a trip to attend the Rio+20 in Brazil in May 2012, in which the President expressed his readiness to the UN Secretary-General to lead HLPEP. The President then appointed the head of UKP4 as the chairman of the National Committee for HLPEP to help him in formulating the agenda and Indonesia's position in leading HLPEP. The REDD+ task force at that time becomes the think tank in drafting the HLPEP final report, which was then submitted to the UN Secretary-General in September 2013 in New York. This effort is very important to attach the REDD+ agenda to the sustainable development agenda.

There is also an initiative of the New Climate Economy consisting of seven countries, which was initiated by Norway, Sweden, and the UK. There is also the Tropical Forest Alliance 2020, which is an initiative from the US government and Consumer Goods Forum (3000 companies whose business lines are related to forest).

The Norwegian government also actively urged the commodity producers and consumers in various countries to commit to zero deforestation at the Climate Summit in September 2014 to the companies such as Asia Pulp and Paper, APRIL Group (logging company) and Wilmar International (palm oil plantation). However, significant impacts of this commitment have yet to be realised until now.

The REDD+ concept that was proposed by Indonesia is now able to influence the concept used at the international level. However, the REDD+ implementation in Indonesia itself is very difficult. Hence, only part of the international commitment and support can be realised. Until the end of 2014, Indonesia has only utilised USD 40 million of the total commitment of USD 1 billion from Norway (4%).

The National Interactions in REDD+ Dynamics

Some of the authors participated to direct the REDD+ task force in 2010. Our effort has successfully created some innovation in forest and land management in Indonesia. Improving forest and land governance is a prerequisite to implement beyond carbon and more than forest REDD+. Most of the REDD+ task force members are new players who suddenly have to deal with other actors such as the Ministry of Forestry (MoF), Ministry of Environment (MoE) and National Council on Climate Change (DNPI), which have grappled with this issue for a long time. Some of the old actors are in doubt with the REDD+ task force members' experience because the REDD+ task force's new members have no bad experience in the past; they act freely, fast and flexibly. These major changes were not anticipated by the former actors and cause a lot of resistance.

The constellation of national actors in the REDD+ issue is very interesting. At that time, one of the authors was entrusted to manage and to create innovation in REDD+. Unfortunately, forestry and other land sector issues, which are very related to REDD+, are still entrusted to the existing ministries and agencies. The climate change issues that are related to REDD+ are also controlled by the DNPI. In addition the land use issues are also related to the interest of palm oil and paper industries. Since all of these related issues are managed independently by various entities, the implementation of REDD+ was hampered.

The President actually hoped that the REDD+ task force will be able to solve the current forest and land governance problems. In one of the cabinet meetings, one of the authors was asked to present the difference between forest area calculations based on different maps produced by the MoF and the MoE. This meeting concluded that one reference map is very important. The President then asks for an immediate moratorium instruction that led to the One Map movement. Unfortunately, this instruction was not fully supported at the implementation phase.

In addition to the moratorium and the one map problems, there is another problem between the government and the indigenous peoples. The Indonesian constitution recognises the rights of indigenous people. Unfortunately, the forestry law considers the indigenous forests as state forests. Forest utilisation licences granted by the MoF sometimes do not take into account the forest area that belongs to the indigenous peoples. However, the situation changed after the Constitutional Court Decision No. 35 in 2013. This decision excludes the indigenous forests from the state forest and therefore entitles the indigenous peoples to control and manage their forests. The REDD+ task force that wanted the recognition of indigenous rights so that a sustainable forest and prosperous society can be achieved then supports this decision. The REDD+ task force then facilitates the indigenous peoples to gain state recognition. On the other hand, the MoF avoided dealing with this matter, handing it over to the local governments. The uncertainties about the recognition and protection of indigenous peoples continue to occur. The REDD + task force and later the BP REDD+ tried to take over this responsibility. This has made the REDD+ task force and BP REDD+ seen as a channel between indigenous peoples and the government.

As the head of the UKP4, one of the authors plays an active role in coordinating related ministries and agencies to overcome these problems. This effort was done by establishing a national inter-sectoral coordination mechanism. Through these efforts, a common understanding with various institutions can be established such as the Ministry of Welfare, Ministry for Political, Legal and Security Affairs, MoE and various NGOs such as Indigenous Peoples Alliance (AMAN), Greenpeace, Indonesian Environmental Forum (WALHI), etc.

The author only asks decision from the President for really critical matters, for example, when two presidential decrees are needed, i.e. a decree to establish BP REDD+ and a decree about a national REDD+ strategy. It is important to maintain the focus on the resulting decisions. Because the establishment of BP REDD+ is one part of the agreements in the LoI, the author prioritises this objective. To pursue this objective was not easy, especially because the author demands that these agencies should be at the same level with other ministries, as mentioned in the LoI.

The national REDD+ strategy is a picture of how the REDD+ task force considers REDD+ not only as the climate change agenda but also as reformation agenda, land-based management of natural resources, and sustainable development agenda that focuses on improving the welfare of local and indigenous communities. This document was prepared by involving various experts and stakeholders, making it a complete document and in accordance with the needs in the future development. The national strategy document has become the core of BP REDD+ preparation and future actions. Unfortunately, this document still cannot convince the ministries to be used in their implementation agenda.

The dynamics that gave birth to BP REDD+ clearly illustrates the competition between actors. Although the President has agreed that BP REDD+ will become an institution at ministry level in 2001, the realisation of this agreement takes 1 year (August 2012–August 2013). This was caused by a lot of disagreements on the BP REDD+ draft. These disagreements occurred because many of the proposed BP REDD+ authorities (especially fund management and supervision) clashed with the existing authorities of other ministries and agencies. As a result, the discussion of the draft, led by the Coordinating Minister for Politics and Law, was prolonged until the end of 2012. Nevertheless, the Minister for Politics at that time could facilitate this discussion and maintain the atmosphere of the debate. The most contentious points include (i) ministry-level status, (ii) roles and functions that were allegedly overlapped with many ministries, (iii) nonstructural entity status, (iv) the financial authority and (v) the monitoring, reporting and verification (MRV) authority. With the help of one of the authors in facilitating communication between the heads of relevant institutions, the final draft to be submitted to the President was finally agreed upon. The decree on BP REDD+ was finally signed by the President on August 31, 2013.

The Local Interactions in REDD+ Dynamics

REDD+ is a global initiative that was boldly adopted as a national agenda and then was operationalised in an innovative way into the local context. The Central Kalimantan province was defined by the President as a pilot province and became the first living laboratory for the REDD+ task force. The REDD+ operationalisation in Central Kalimantan was not an easy task. This was the first time the REDD+ was tested on a provincial scale that has a different political context. At that time the Governor and many of the mayors in this province came from the opposition parties. Nevertheless, the Governor has a good communication with the President, so the REDD+ concept was easily accepted. This is because the Governor saw a lot of potential benefits that can be obtained from the REDD+ implementation.

However, many difficulties arose in implementing the REDD+ concept in Central Kalimantan. The problems faced by the REDD+ task force were (i) high deforestation rates (100,000 ha/year), (ii) a lot of unproductive and wastelands, (iii) poor spatial planning, (iv) the threat of forest clearance for mining, (v) the local governments having only limited manpower and (vi) natural resource corruption cases. The good leadership from the Governor was not supported with adequate human resources in Central Kalimantan. To overcome this, the REDD+ task force then established a support office in the Governor's office. This support office serves as the secretariat for the Regional REDD+ Commission that runs REDD+ activities in Central Kalimantan.

Various programmes had been tested in Central Kalimantan. Among others, the Governor offered to test the REDD+ programme through a rehabilitation project of an abandoned one million hectares of peatland project area (a failed project in 1996). A number of innovations are attempted such as green villages, green schools, indigenous peoples mapping, the prevention of forest fires and land use licensing improvements. The involvement of local stakeholders is very essential in determining the success of such programmes. A lesson learned from this pilot province is that the operationalisation of REDD+ should be done in an innovative way through real programmes that can be directly linked to the local needs.

The Governor cannot act by himself in managing REDD+, and the involvement of district mayors is also necessary. The decentralisation policy has given many implementation authorities to the district mayors. At that time, the Governor faced many difficulties in controlling all of the 15 district mayors. Only seven district mayors, who have the same political alignment with the Governor, declared their readiness to support REDD+. Therefore, the BP REDD+ tried to convince the other mayors to support this concept. The strategy chosen by the BP REDD+ was to appoint the Governor as the regional leader of REDD+ to coordinate the REDD+ implementation at the district level. The new Regional Government Law No. 23 of 2014, which gives greater powers to the Governor, is used as the basis in bargaining the position of the Governor.

The synergy between the Governor's and the mayors' leadership is very detrimental to the success of the REDD+ operationalisation. The REDD+ strategies at the local level should match with the vision/mission and strategy at the provincial and district levels. Our experience has shown that without it, the political support from the local government could not be obtained. The various facilities built by the REDD+ task force eventually had to be maintained and operationalised by the local government. That is why the coordination is very important from the outset.

After BP REDD+ was established, the REDD+ activities were expanded to ten other provinces, i.e. Aceh, West Sumatra, Riau, Jambi, South Sumatra, West Kalimantan, East Kalimantan, Southeast Sulawesi, Papua and West Papua. The dynamics in each province is different, making the BP REDD+ operationalisation context of REDD+ vary in each province. The leader's character also determines the structure of REDD+ in each province. The Governor in Central Kalimantan wanted to lead this process directly (top-down), while the Governor in Jambi saw a great need to involve all mayors (bottom-up). The BP REDD+ operationalisation also appears to be more difficult in Papua than in Sumatra and Kalimantan. In Papua, the human resource capacity issues became the main constraint in REDD+ operationalisation. The strategy and approaches are determined depending on the needs of the given province and as a result of communication and coordination with local stakeholders.

The BP REDD+ introduced ten imperative programmes to be implemented at the provincial level: (1) prevention of forest fires, (2) identification and protection of indigenous peoples, (3) structuring land use licensing, (4) land conflict resolution, (5) moratorium monitoring, (6) green villages, (7) green schools, (8) law enforcement improvement, (9) spatial planning improvement and (10) national park development. None of these programmes directly related to the emission reduction. These programmes are designed to overcome the causes of deforestation and land degradation and more relevantly to improve forest and land use governance as well as to empower local communities. These REDD+ implementation activities at the local level were designed to be concrete and in accordance with the local needs. Emission reduction programmes can only be carried out if the previous programmes could be well received by the local communities.

A needs-oriented approach became the focus of BP REDD+ at the end of 2014. The Riau province, for example, was hit by forest and land fires throughout 2014. Hence, BP REDD+ came up with forest and land fire prevention programme. The BP REDD+ acted in collaboration with the Riau province government to (i) develop a forest and land fire monitoring system, (ii) conduct an audit of compliance with

17 companies and (iii) develop local government capacity. The last action was taken seriously by the provincial government which then leads to a major action plan to improve forest and land governance in the Riau province. The needs of the West Sumatra province are different. The governors and the local government wanted to develop forest-based villages. There were 500,000 hectares of forest to be proposed to get a village's forest status from the government. By empowering village communities in protecting the forests (green village), the BP REDD + approach was in line with the needs of the local governments in obtaining a village's forest status. While in Jambi, according to the needs of the province, BP REDD+ focused on the empowerment of indigenous peoples. The needs-based approach was successful to ensure that the BP REDD+ agenda can be implemented at the local level.

The role of NGOs is also very detrimental in REDD+ operationalisation. In Jambi and West Sumatra provinces, for example, the collaboration of BP REDD+ and the Conservation Information Centre (WARSI, an NGO that is based in those provinces since 1992) has made it easily accepted by the local communities. In Papua, the BP REDD+ activities are supported by the World Wide Fund for Nature (WWF) and the Samdhana Institute, while in other provinces, where the NGOs do not actively participate, the operationalisation of REDD+ becomes much more difficult. Therefore, it can be concluded that to support a national agenda such as REDD+, the active participation of the NGOs is very important.

The innovations made by the BP REDD+ that have been mentioned previously may improve the bargaining power of Indonesia at the international level. The success stories of various REDD+ implementations may attract more support from the international community. Unfortunately, the existence of BP REDD+ is still vulnerable to the dynamics of the national politics, i.e. regime change. This dynamics may also give negative influence to the existing REDD+ programmes.

Lesson Learned

This paper presents a case study of development policy at the national level using the topic of REDD+ implementation. Although this case occurred in Indonesia, the dynamics in this case involves local, national and international actors.

From this case, there are several lessons that can be learned. First, the process undertaken by the REDD+ task force to develop BP REDD+ was started from intellectual design to establish the desired condition (sustainable development beyond carbon and more than forest). This process was done by involving various stakeholders beyond the REDD+ task force. It was continued with a dynamic and iterative process that involved information gathering from various sources, intensive alternative design by the REDD+ task force, extensive stakeholder consultation at provincial and national levels and cross-ministry and agency debate. This process produced various outputs, such as the REDD+ national strategy, a funding instrument, and various programmes at provincial and district levels. This action research showed that the policy development cycle, which has been discussed in previous literature, can be applied at the national level.

These collaborative campaigns are expected to be able to create a paradigm shift and establish a significant foundation to make these initiatives sustainable. The 10-year target (2010–2020) clearly indicates that this process will experience regime change and transition. This is a challenge, because the structure and the working process of the government often changes between regimes.

References

Ackoff, R. L. (1974). *Redesigning the future: A systems approach to societal planning*. New York: Wiley.

Ackoff, R. L. (2001). A brief guide to interactive planning and idealized design. Retrieved on March, 19, 2006.

Dyson, R. G., Bryant, J., Morecroft, J., & O'Brien, F. (2007). The strategic development process. In *Supporting strategy: Frameworks, methods and models*. Chichester: Wiley.

Friend, J. K., & Hickling, A. (2005). *Planning under pressure: The strategic choice approach*. Amsterdam/Boston: Elsevier/Butterworth Heinemann.

Jann, W., & Wegrich, K. (2007). Theories of the policy cycle. In F. Fischer, G. J. Miller, & M. S. Sidney (Eds.), *Handbook of public policy analysis: Theory, politics, and methods*. Boca Raton: CRC Press.

Jones, B. D. (2002). Bounded rationality and public policy: Herbert A. Simon and the decision foundation of collective choice. *Policy Sciences, 35*, 269.

Milano, M., O'Sullivan, B., & Gavanelli, M. (2014). Sustainable policy making: A strategic challenge for artificial intelligence. *AI Magazine, 35*, 22–35.

Pidd, M. (2004). Contemporary OR/MS in strategy development and policy-making: Some reflections. *The Journal of the Operational Research Society, 55*, 791–800.

Rao, V. S. P., Krishna, V. S. P. R. V. H., & Krishna, H. V. (2004). *Strategic management*. New Delhi: Books Excel.

Simon, H. A. (1986). Rationality in psychology and economics. *Journal of Business, 59*, S209–S224.

Sterman, J. D. (2001). System dynamics modeling. *California Management Review, 43*, 8–25.

Value Co-creation Platform as Part of an Integrative Group Model-Building Process in Policy Development in Indonesia

Utomo Sarjono Putro

Abstract The Indonesian government nowadays faces multidimensional and multifaceted problems that require a more proper framework of policy making. Group model building (GMB), in parallel with system dynamics, becomes a successful methodology in policy making though some weaknesses exist. We have proposed an integrative GMB process by adapting the framework of value co-creation and orchestration platform to overcome those problems. Three management strategies of value co-creation as the main findings are revoked: involvement, curation, and empowerment. Those keywords are inputted to improve the GMB process and applied to the wicked problems found in the Indonesian context that needs more integrative solution to see how the GMB platform may be implemented in the real-world problem. Although it is still an ongoing process, we can see that by adapting the value co-creation platform as part of the GMB process, it is advantageous in defining what activities should be done by each actor (e.g., facilitator, modeler, stakeholders) in order to have a better policy recommendation in the part of policy making and policy development process.

Introduction

The Indonesian government is progressively being tasked with solving complex policy issues. Some of these policy issues are so complex they have been considered as wicked problems. The term "wicked problems" is found in many practices, including public administration, health education, ecology, forestry, and most often in policy science (Batie 2008). It can be characterized as a problem that makes it hard to clarify the objectives that want to be achieved, which involves many stakeholders and involves different versions of the problem that is currently faced (Australian Public Service Commission 2007).

In Indonesia, wicked problems have been appearing with higher frequency on the policy agenda, e.g., climate change, corruption, unemployment, national debts, recessions, ecological degradation, water and food shortages, lack of health care,

U.S. Putro (✉)
School of Business and Management, Institut Teknologi Bandung, West Java, Indonesia
e-mail: utomo@sbm-itb.ac.id

© Springer Japan 2016   17
K. Mangkusubroto et al. (eds.), *Systems Science for Complex Policy Making*,
Translational Systems Sciences 3, DOI 10.1007/978-4-431-55273-4_2

and so on (Roberts 2012). Although the Indonesian government has worked on efforts to handle these problems, many agreed on the fact that wicked problems were indeed unmanageable. Moreover, the current conditions that need improvement are each of the actors and stakeholders in Indonesia still has difficulties to bridge each need and to interact with others. There is a need of an integrated process in solving the problem between government and other stakeholders, such as local governments, local communities, industries, and academic communities, to have a better policy development.

One of the methodologies that are successful in the content of policy making globally is the group model building (GMB) using system dynamics. GMB is a well-known process in policy making which includes many interactive processes before building and executing the model. GMB is considered to be an appropriate process to define a problem where there are many stakeholders included in the problem or when the problem is too complicated to be tackled. However, previous literature has stated that some limitations can be found in the GMB process. First, the context in which group modeling projects take place may be important (Vennix 1999). Moreover, factors such as the type of organization, the organizational culture, and the history of participants may affect the project implementation (Bérard 2010).

We proposed an integrative group model-building process by adapting the framework of value co-creation and orchestration platform in order to enhance the process of policy making and policy development in Indonesia, considering some limitation on the character of policy making in Indonesia that needs more integrative solution. Moreover, by extending the process using the platform, it is expected that the process will give greater advantage, which is creating more value in the process of policy making and policy development.

The rest of this chapter is presented by using a systematic writing organization. First, the background on GMB and why its process can be extended will be briefly explained. It will be followed by the concepts of value co-creation and orchestration platform. Next, the proposed value co-creation platform on the GMB process will be defined. In addition, the application of the platform that has been implemented in some process in policy making in Indonesia is given. Last but not the least, a conclusion of remarks will be shown in the last section.

Existing Platform for Policy Modeling

Decision-makers design policies that are often difficult to implement, because the design fails to take into account the key feedback that will generate undesired consequences or limit the benefits of actions (Bérard 2010). This may be the case more specifically in circumstances when the system under consideration has many components that may not be easily taken into account by decision-makers' mental models. This fact leads to the need of using approaches such as system dynamics modeling that helps to recognize the dynamic behavior that a system experiences and, consequently, also to mitigate the cognitive limits of decision-makers.

The modeling process using system dynamics can be carried out through two types of projects: modeling projects and group modeling projects. The first type is managed by one or more modelers, who design the models themselves and gain the expertise and required data from many sources and often from experts on the modeled system (Bérard 2010). While the second type of project, group model building, refers to a system dynamics model-building process in which a group is deeply involved in the process of model construction (Vennix 1999).

In a group model building (GMB), the participants develop one or many models during structured sessions with the help of a facilitator, who must favor the clarification of knowledge within the group (Bérard 2010). These activities are typically formed as workshops, work sessions, or conferences. The participants of the workshops are the "clients" for whom the model is developed, and can be researchers that are the expert in the problem being discussed, and/or practitioners who are themselves actors of the system.

There are five components of the process in doing GMB, as mentioned by Bérard (2010): problem articulation, dynamic hypothesis, simulation model formulation, model testing, and formulation of strategies and evaluation. However, each step of the group modeling process may include a succession of individual activities, subgroup workshops, and plenary sessions. It depends on how suitable the activities are to the actors that are involved in the GMB.

Although previous literature has clarified on how the GMB process could be applied to some problems in policy making and policy development, some limitations still should be noted. First, the context in which group modeling projects take place may be important (Vennix 1999). Another suggestion is that factors such as the type of organization, the organizational culture, and the history of participants may affect the project implementation (Bérard 2010). In addition, the model is limited to the methodological frameworks using system dynamics modeling. While in the practice, other methods of modeling could be considered in the process.

To cope with the limitations, we propose a group model building where value co-creation platform becomes a part in the GMB process to enhance the policy making and policy development process. We implement the platform based on the real ongoing case study in Indonesia, considering the fact that the type of factors mentioned (organizational, historical, or cultural type) has a role in influencing the project implementation of the GMB process. In the next session, a brief explanation on value co-creation platform will be presented.

Value Co-creation and Orchestration Platform

Service science defines service as a phenomenon observable in the world in terms of a service system with value co-creation interactions among entities by taking a bird's-eye view of various perspectives in which service system entities can be people, businesses, nonprofits, government agencies, and even cities (Kijima, Rintamki, & Mitronen, Value Orchestration Platform: Model and Strategies, 2013). Co-creation itself is a very broad term with a broad range of applications.

Value co-creation is an emerging concept in business, marketing, management, and many other practices. Over the last few years, there have been researches that provide a starting point in the discussion of different perspectives on value co-creation. The term value co-creation gains much interest in innovation research that requires the adoption of new terminology, frameworks, and fields of research exploration. The potential advantages from the adoption of value co-creation practices and strategies are one main point that needs to be further addressed.

Kijima et al. (2013) has identified value co-creation interaction as an active, creative, and social process based on collaboration between the provider and customer. It is a form of collaborative creativity of customers and providers that is used to enhance the organization's knowledge-acquisition processes by involving the customer in the creation of meaning and value, although it is often initiated by the provider. There are four phases of the value co-creation process (Kijima et al. 2013) which are: co-experience, co-definition, co-elevation, and co-development.

In co-experience, the provider and customer may need to share an internal model to co-define a mutual understanding about the process or the problem that want to be defined. By interacting with each other, the customer and provider may learn about the other's preferences, capabilities, and expectations so that they may co-define and share a common internal model (Galbrun and Kijima 2009).

Co-elevation is a zigzag-shaped spiral up process of expectation of the customers and abilities of the providers. Higher expectations by individuals lead to higher quality and greater value. Moreover, co-development pays attention to co-innovation generated by simultaneous collaboration among the various entities (Novani et al. 2015).

In the value orchestration platform, there are three management strategies to be implemented: involvement, curation, and empowerment. The platform is concerned with the methods to involve appropriate customers and providers in the platform and to vitalize interactions between customers and providers. Curation is essential for the platform to encourage customers and providers to co-elevate and co-develop, while empowerment refers to how a platform empowers customers and providers so that each side finds the other attractive and both are motivated to interact with each other (Kijima et al. 2013). The three management strategies and value co-creation platform framework are shown in Fig. 1.

By adapting the framework of value co-creation and value orchestration platform, we aim to look at how the GMB process can be approached using this platform, where in the process, each actor has an objective to have value co-creation on each of the phase in the GMB process. The proposed model will be defined in the next session.

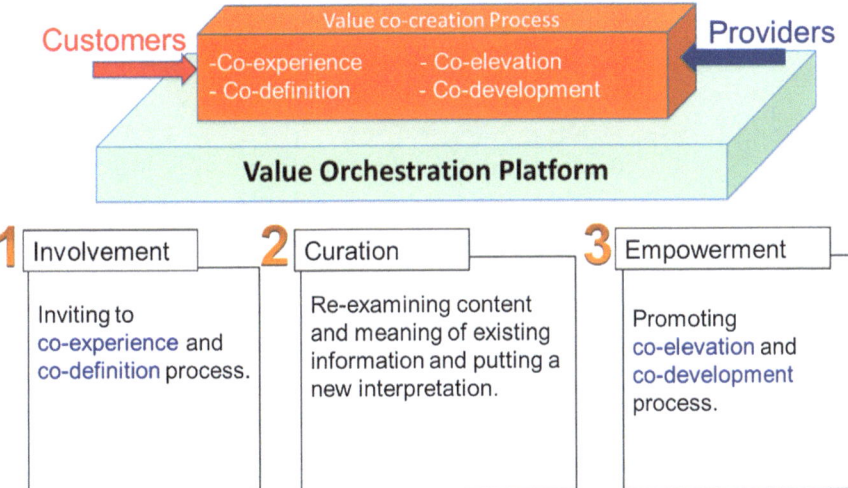

Fig. 1 Three management strategies included in the two-layered service system (Novani et al. 2015; Galbrun and Kijima 2009)

Value Co-creation: Group Model-Building Platform

We proposed an integrative group model-building process by adapting the framework of value co-creation and orchestration platform. *Why*? Because the existing group model building using system dynamics (Vennix 1999) can be seen as having the process of involvement, curation, and empowerment in the process itself. Moreover, by extending the process using the platform, it is expected that the process will give greater advantage, which is creating value, in the process of policy making and policy development.

There are three main actors involved in the proposed GMB platform: the facilitator, stakeholder, and decision-maker. The facilitator is responsible to facilitate the whole activity included in the GMB process. The tasks of the facilitator are inviting relevant stakeholders, decision-makers, and modelers, grouping the issues, making intensive discussion within a sector or among sectors to improve and/or to develop a model that can be used to resolve the issues, and confronting and coordinating the results from the model to assist implementation of the model. The role of the facilitator is considered as most important, where they have to maintain all parties to be engaged to each other and work together to solve problems.

Stakeholders are the actors who are involved in the issues that want to be solved. The tasks of the stakeholder are describing the issues based on some questions such as *What is the current situation happening? What approach has been done to resolve the issues? Has the approach been able to give in-depth understanding to the policy development? What is the main weakness of the approach?* After the issues have been identified, the next tasks are to discuss relevant ideas that can be

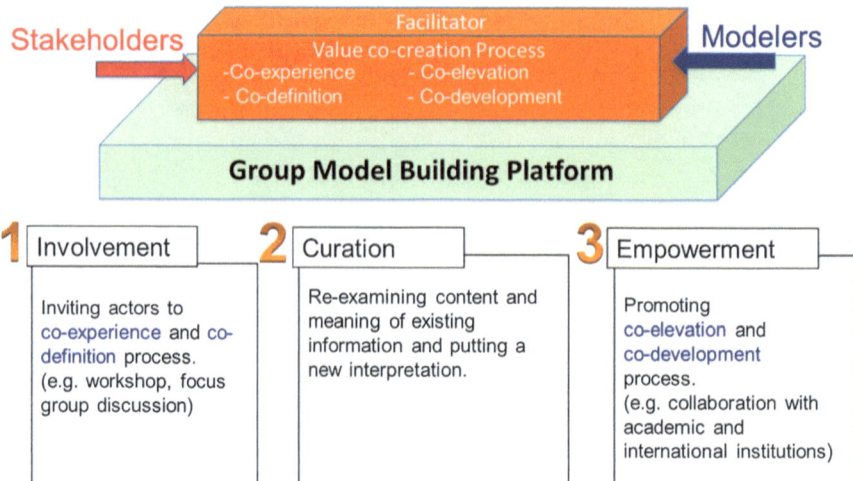

Fig. 2 Proposed group model-building platform and management strategies in two-layered service system

implemented to improve the quality of solving the problem, making a model together with the modeler, and analyzing feasibility of the results to be implemented.

The third actor is the decision-maker. The decision-maker is responsible to decide what kind of policy resulted from the process of GMB. The tasks of the decision-maker are making confirmation about the current issues and the previous policy that has been implemented, getting involved in discussion and the process on developing the model together with the modeler, and analyzing the feasibility of the results to be implemented.

To generate the integrative strategies in the GMB platform shown in Fig. 2, we have implemented the three management strategies in building a value co-creation platform according to Galbrun and Kijima (2009). The three management strategies in the GMB platform for policy making development are as follows:

- *Involvement*. Inviting all actors so that co-experience and co-definition process could be accomplished. The platform orchestrator (facilitator) is primarily having the focus in how to get the stakeholders "on board" to involve in the platform and to vitalize interactions between the customer (stakeholder) and provider (modeler). This involvement strategy can be implemented through some events such as workshop and formal focus group discussion, where all three actors are in the same board at the same time.
- *Curation*. Value curation is essential for the platform to encourage all actors to co-elevate and co-develop. The curation strategy can be implemented by reexamining content and meaning of existing information of the problem and putting a new interpretation in looking at the problem. In this process, actors together have to collect, select, analyze, edit, and reexamine content and

meaning of existing issues in order to put a new interpretation on and give a new meaning to them. Based on the newly developed interpretation and meaning, value co-creation process is expected to be improved.

- *Empowerment.* Empowerment is another aspect of value orchestration, particularly for the co-elevation and co-development phases (Novani et al. 2015). Promoting co-elevation and co-development process could be done when stakeholders are empowered by lifting up their aspiration level. In this process, they could consider the effort through collaboration with international institutions, establishing group model-building activities on priority sectors, and capacity building for policy makers.

The integrative GMB platform is shown in Fig. 3. This whole process is advantageous in enhancing the value co-creation among each stakeholder and actor in the development of policy making. Once the relevant stakeholders were invited, interaction between stakeholders is conducted. The literature review aims to identify theoretical models that already exist or are being developed by academics or previous works and literature. Moreover, it is the phase where all actors identify the models that are available and can be used as input in the next phase. Based on the initial models, relevant stakeholders are invited to the workshop (WS1, WS2, and WS3). In the model-building and validation phases, the stakeholders that are potentially concerned and interested in evaluating and developing

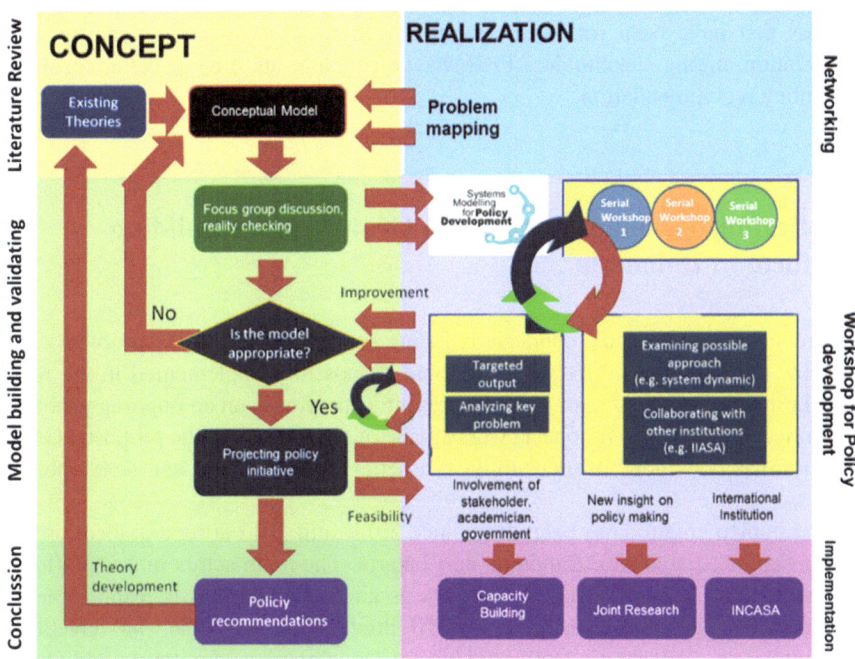

Fig. 3 Proposed value co-creation platform in the GMB process

the initial models are involved. These first two phases are considered as the starting process of *involvement* in the three management strategies previously discussed.

Using this interactive GMB process, the existing model which has resulted from the literature review can be examined and adapted into the model-building process. The modeler, facilitated by the facilitator, however, should use a common language where all the actors involved could understand in order to get a new interpretation on the problem and on how the problems can be solved.

Each stakeholder may express their assumptions and perceptions, so the process will be more comprehensive and a robust model can be produced. At this phase, stakeholders may work together with the modeler, so that the initial model from academicians could be criticized, modified, and adjusted based on their needs and knowledge. In the process of model building and validating, value *curation* is preceded by the activities of reexamining the previous model and contrasting it with the real-world problem where all stakeholders and modelers discuss and try to build a more suitable model for the current problem.

In addition, in the model-building and validating process, value *empowerment* could occur by having a collaboration with institutions such as the government, academic institution, and international institutions that also have a concern on the problems that are being discussed. With this collaboration, it is expected for the stakeholders to be more empowered so that they can lift up their aspiration level to solve the current problem.

In the conclusion part of the integrative GMB process, each stakeholder may propose some recommendations to the policy making process based on the whole process that have been followed. Hopefully, there is consistency, synergy, and cooperation among stakeholders to solve the problem and give a better result in the policy recommendation.

Application of Value Co-creation Group Model-Building Platform in Indonesia

As previously mentioned, Indonesia is facing some complex policy problems. To see how the proposed GMB platform could be possibly implemented in the real-world problem, in this section we will present a study case on an ongoing problem of the rice production in Indonesia where some of the process on the proposed GMB platform has been done in order to have a better policy making and development process.

The implementation has been done since 2011 until 2012 at several locations in Indonesia, i.e., Jogjakarta, Bandung, and Jakarta. The main actors in this platform are facilitators. We define the facilitators as an institution that is responsible to organize, arrange, and manage the GMB process. In this case, we have two facilitators, i.e., President's Delivery Unit for Development Monitoring and Oversight (*UKP-PPP*) of the Republic of Indonesia and the School of Business and

Management (*SBM-ITB*), Bandung Institute of Technology. *UKP-PPP* is a governmental institution that is responsible to invite relevant stakeholders in the governmental level, i.e., ministries, agencies, and local governments, and to facilitate fund to hold the workshop and seminar. *SBM-ITB* is a public educational institution that is responsible to invite relevant stakeholders in the academic level, in this case as the representative of universities and academic/practitioner in a local and international scope.

At the literature review phase, researchers from academic institutions have several approaches for policy development in Indonesia, i.e., people-centered approaches, the application of economic models for environmental policy development, system dynamics approach, social network analysis, and geospatial modeling of national economic corridors. In this step, the researchers from academic institutions make a mapping of issues and the approaches as well.

The process is continuing to the discussion of issues and the existing approach. A series of workshops involving academia, industry, and government (e.g., ministries, agencies, local governments) were planned and conducted. Three workshops have been done in 4 months starting on 22 October 2011, 26 November 2011, and 28 January 2011. These three workshops involved 24 presenters, 6 invited speakers, and 111 submitted papers and were attended by 226 participants, which consist of 38 % academics, 14 % practitioners from private companies, 17 % practitioners from state-owned research and development agencies, and 31 % representatives of the government (e.g., ministries, presidential office). The intensive discussion was conducted to resolve the issues from each sector. An academic facilitator is provided to manage the discussion. In the discussion, one of the presenters explains the issues and methodology he/she uses to tackle the problem. After that, the stakeholders are involved to give their opinion and critics based on their own experience.

After several workshops in 2011, the facilitators evaluated the results of the workshops and identify special or crucial issues that should be handled. To promote the issues, the facilitators conducted thematic workshops on priority sectors. The workshop is aimed at confronting and coordinating the results from the previous discussion about the issues and the methodologies in a more intensive way. Thematic workshops were conducted to closely connect academics, industry, and policy makers in formulating policies in those specific sectors. The first thematic was run on 25 February 2012 focusing on food security with the theme *Achieving 10 Million Ton Rice Production Surplus in 2014*, which involved relevant ministries (e.g., Ministry of Agriculture, Ministry of Public Works, Ministry of Trade, and Ministry of State-Owned Enterprises), agencies, state-owned enterprises, academics, and other related stakeholders. The first thematic workshop discussed the strategies, requirements, and challenges to achieve the 10 million ton surplus.

Through the ongoing processes, we have implemented the proposed management strategies, i.e., involvement, curation, and empowerment in the activity.

- *Involvement*: having a focus group discussion and several workshops to have the co-experience and co-definition process on the issues that have been discussed.

The facilitators have chosen and invited relevant stakeholders to be "on board" and facilitate the interactions to be going on between them. This involvement strategy also involves:

1. Developing research networks and national stakeholders: In Indonesia, there have been many researchers and research products that can be related to national and regional policies. On the other hand, there are many stakeholders from the world of business and government who need policy analysis inputs. Group model-building activities are started by developing a database of these researchers and stakeholder networks. The stakeholders who have been involved in this activity are the Ministry of Energy and Mineral Resources, Ministry of Research and Technology, Ministry of Agriculture, Ministry of Health, etc.
2. Developing a network of international scientists and stakeholders: Not only in Indonesia, many international research institutes have excellent policy analysis capacity, for example, the International Energy Agency (IEA) which includes Indonesia as one of the countries in their energy analysis. In addition, there is also the International Institute for Applied Systems Analysis (IIASA), which is a world-renowned institution in the global policy analysis based on the system analysis. In establishing this international network, Indonesia became a member of IIASA in June 2012, by first establishing the Indonesian National Committee for Applied Systems Analysis (INCASA) as a national member organization of IIASA.

- *Curation*: within the thematic workshops, reexamining content and meaning of existing information and putting a new interpretation are accomplished. The facilitators have to be proactive to select some crucial issues to follow up. One of the implementations of this strategy is to activate a *triple helix* interaction. The *triple helix* interaction between academics, business, and government is an important part of group-based modeling for national policy making. In achieving the triple helix interaction, the workshop format is considered as the most appropriate format. With the workshop, researchers can directly receive feedback and input from policy makers in business and government. Thus, researchers can directly improve their research, to accommodate different perceptions expressed by workshop participants.
- *Empowerment*: the facilitators have conducted some collaboration with international institutions to promote co-elevation and co-development process. Some implementation of this strategy includes:

1. Capacity building: Though several inputs for the policy makers have been obtained through workshops, the ability of policy makers to analyze a system or a problem needs to be developed. The increase of their capacity will also contribute significantly to the process of the future GMB process. Policy makers in the future are expected to make necessary revisions directly to research and the policy by using GMB. Some trainings and tutorials have been given with some cooperation with the ministries and agencies. System

dynamics, one of the approaches which is considered being the most general and understandable approach, is used to be one approach in the capacity building process.

2. Joint research: GMB activity followed with joint research activities as well. Joint research is an attempt to solve the actual problems facing policy makers today. In Indonesia, some joint research activities are still on the process, including policy analysis together with IIASA as an international institution. On the food sector, to illustrate the process of group-based modeling has been carried out; the following example was appointed joint research processes related to the development of policy making in the beef industry in West Sumatra. SBM-ITB, as an academic institution, was asked to participate in the capacity building process and the modeling of the system. The case also involves government institutions (regional and national). With the collaboration between international institutions, academic institutions, and government institutions, an integrative GMB process could be advantageous in the policy making and development process (Fig. 4).

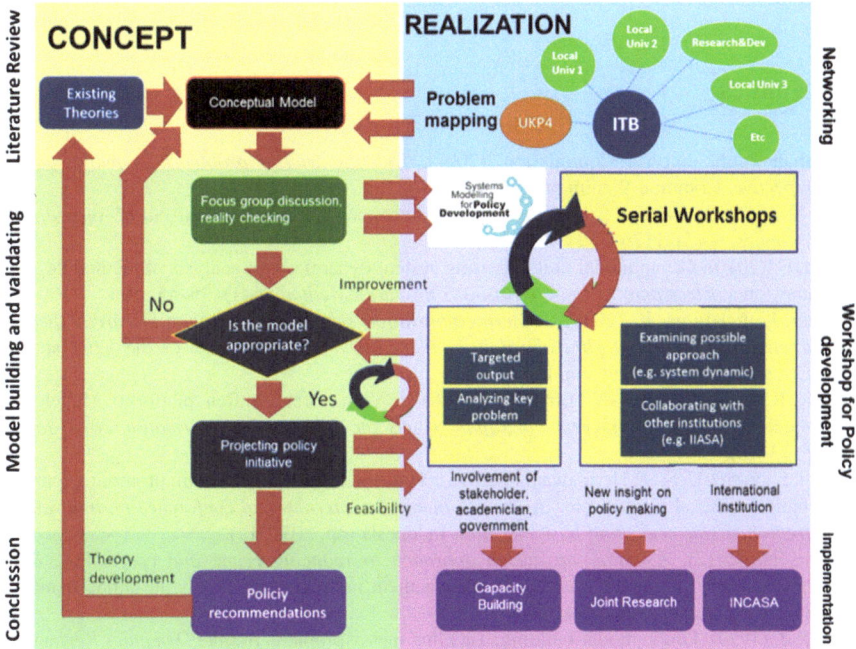

Fig. 4 Proposed value co-creation platform in the GMB process on the case study in Indonesia

Conclusion

We have proposed an integrative group model-building process by adapting the framework of value co-creation and orchestration platform in order to enhance the process of policy making and policy development in Indonesia, considering some limitation on the character of policy making in Indonesia that needs more integrative solution. By extending the process using the platform, the process of GMB in terms of policy making and development will give greater advantage, which is creating more value for all actors that participate in the process.

In the proposed GMB platform, we can define three management strategies of the value co-creation platform, involvement, curation, and empowerment, which would be the part of the group model-building process. A study case of an ongoing problem in Indonesia was presented to see how the GMB platform may be implemented in the real-world problem. Though it is still an ongoing process, we can see that by adapting the value co-creation platform as part of the GMB process, it is advantageous in defining what activities should be done by each actor (i.e., facilitator, modeler, stakeholder) in order to have a better policy recommendation in the part of policy making and policy development process.

References

Australian Public Service Commission. (2007). *Tackling wicked problems: A public policy perspective.* Common Wealth of Australia.
Batie, S. S. (2008). Wicked problems and applied economics. *American Journal of Agricultural Economics, 90*(5), 1176–1191.
Bérard, C. (2010). Group model building using system dynamics: An analysis of methodological frameworks. *Electronic Journal of Business Research Methods, 8*(1), 35–45.
Galbrun, J., & Kijima, K. (2009). *Fostering innovation system of a firm with hierarchy theory: Narratives on emergent clinical solutions in healthcare.* Proceedings of the 52nd annual meeting of the ISSS, Madison, USA.
Kijima, K., Rintamki, T., & Mitronen, L. (2013). Value orchestration platform: Model and strategies. In *Annual conference of Japan society for management information 2014, Japan* (pp. 13–16).
Novani, S., Putro, U. S., & Hermawan, P. (2015). Value orchestration platform: promoting tourism in batik industrial cluster solo. In *The 6th Indonesia international conference on innovation, entrepreneurship and small business, Bali,* Indonesia (pp. 207–216).
Roberts, N. (2012). *A bottom-up design approach to reducing crime and corruption.* 2012 conference of the international public management network, innovations in public management for combating corruption, Hawaii.
Vennix, J. (1999). Group model-building: Tackling messy problem. *System Dynamics Review, 15*(4), 379–401.

The Effect of Sharing Information on Learning Process in Experimental Duopoly Game: A Case from Indonesia

Manahan Siallagan and Shimaditya Nuraeni

Abstract This article discusses the implementation of laboratory duopoly market in Indonesia. Some uniqueness of Indonesian context was seen during the experiment, e.g. the respondents tend to ignore the importance of sharing the result of flipped coin which may reduce the uncertainty of the outcome.

However, this result is based on post-graduate respondents only. Some various respondents may provide different results.

Introduction

The research on learning and the availability of information in duopoly market have attracted a great deal of attention. One of interesting questions is "What are the factors affecting competition and collusion in experimental markets?" Related to collusion, the study showed that collusion occurred only when subjects were matched in fixed groups for the entire experiments. If the subjects are matched in random matching, the collusion does not occur and the Cournot–Nash solution is a good prediction (Holt 1985; Huck et al. 2001). Another source of collusion is sharing information. Pre-play communication between two players and announcements of information about non-binding price announcements in posted-offer can make the markets significantly more collusive (Hardstad et al. 1998; Davis and Holt 1993; Cason and Davis 1995). The process of information sharing within the experiment improves the collusive behavior, which is affected by experience of the subjects (Bensen and Faminow 1988). Through the experiment, the subjects tend to use their past experience to determine their next decision. This process is similar to learning process.

The investigation of learning process and information sharing is to examine the rules of learning on the equilibrium selection in oligopoly or duopoly market. The experiments were designed to test various learning theories in the context of a

M. Siallagan (✉) • S. Nuraeni
School of Business and Management, Institut Teknologi Bandung, West Java, Indonesia
e-mail: manahan@sbm-itb.ac.id; shimaditya@sbm-itb.ac.id

© Springer Japan 2016
K. Mangkusubroto et al. (eds.), *Systems Science for Complex Policy Making*,
Translational Systems Sciences 3, DOI 10.1007/978-4-431-55273-4_3

Cournot oligopoly (Huck et al. 1999; Lupi and Sbriglia 2003). It analyzed the relationship between learning theories and the availability of information. The results showed that the availability of information and the rule of imitation could lead to the selection of competitive equilibrium. More information about demand and cost conditions brought in less competitive behavior, while more information about the quantities and profits of other firms yielded more competitive behavior. These results are also in line with Altavilla et al. (2006) and Vega-Redondo (1997), which state that imitation behavior will increase the level of competition in the market. However, information on the industries' average profitability might induce more collusive outcomes. In this sense, the firms perceive the industries' average profitability as aspiration level. The firms will try new strategies whenever their profits fall below the industry's average profitability (Dixon 2000; Dixon et al. 2006).

The availability and learning process can have different outcomes in the duopoly experimental game. The subject tends to use the available information together with his/her experience to make next decision. However, the way the subject makes a decision is not in term of rational decision-making. The subject only uses his/her past experience and it is based on his/her expected goal. He/she main costs were quadratic and represented increasing returns to scale in order to enhance the expected payoff differences for the two information sharing choices, as discussed in Section II. This type of learning process has known as aspiration rule that has a long historical tradition (Simon 1955; Simon 1956). In 1999, Timothy N. Cason creates a laboratory for duopoly markets that examines the importance of information sharing in facilitating tacit collusion under conditions of demand uncertainty. The motive behind the experiment is that economist has a great believe that enhanced information about demand conditions rivals' action can play an important role in facilitating collusion. Unfortunately, the empirical work based on field data is challenging because the information different decision-makers possess and the corresponding residual uncertainty they face are typically difficult to identify.

In this research, we try to investigate the effect of sharing information on learning process by making experimental study. The purpose of this study is to understand what kind of behavior will be emerged in a market of duopoly experimental game. Willingness of the subject to share information will determine the behavior of the market. Through this experimentation, the subject will use his/her experience to determine his/her decision-making process. This research will also explore the effect of subject's aspiration level in the learning process and in the decision-making process.

The Experimentation

The experimentation is based on Timothy N. C and Charles F. Mason (1999) with some modifications. Actually, the original experimentation is computer-based, where the subject plays a repeated Cournot competition game against the same rival throughout the session via computer. However, in this research, we use manual experimentation by introducing some modifications, such as aspiration level and round-robin play. In the round-robin play, a subject will play the game with five persons in one round. Each round consists of five sub-rounds. The experimentation consists of two stages. The first stage is aspiration level stage. In this stage, each subject determines the goal of what he/she expects to achieve in current round. The second stage is an information-sharing stage. In the information-sharing stage, all subjects face two tasks in each sub-round. In task 1, they choose whether or not to share a "coin flip" signal with their rival. After the task 1, the subjects learn their rival's sharing decision. Both signals are revealed to both subjects, only if both of them agree to share information; otherwise, the subjects only learn their own signal (and so are imperfectly informed about demand). In task 2, they make what was framed as an "integer choice" that is corresponded to a quantity choice between 0 and 8. In the task 1, subject will flip a coin, which has two outcomes: Head (H) or Tail (T). Based on this result, a payoff table will be used in the sub-rounds. There are three kind of payoff tables used, Green, Blue, and Red. Table 1 shows the possible outcomes with the payoff table.

When subject A's flipped coin result is Head (H) and subject B's flipped coin result is H, then in this current sub-rounds, both of them will use Green payoff table and the table will be given to them if they choose to share information about the flipped coin. When subject A's flipped coin result is Tail (T) and subject B's flipped coin result is T, then in this current sub-round, they will use RED payoff table and the table will be given to them if they choose to share information about the flipped coin.

When subject A's flipped coin is H (or T) and subject B's flipped coin is T (or H), then in this current sub-rounds, they will use BLUE payoff table and the table will be given to them if they choose to share information about the flipped coin. For all condition, if one of the subjects decide not to share the result of flipped coin, there will be no information about what payoff table will be given. However, the result outcome will use the payoff table according to the coin-flipped outcome (only the instructor know what type of payoff table used for calculation).

In this experiment, we use truncated demand for payoff matrices. For green payoff, the symmetric static Nash equilibrium is 180, for blue is 63, and for red is 0. There is no symmetric joint profit maximum in those of payoffs. As we can see in Tables 2, 3, and 4, the green payoff table has positive outcome compared to the other payoff tables.

Table 1 Possible outcomes of payoff table

	Subject A' flip coin	
Subject B flip coin	HEADS	TAILS
HEADS	GREEN	BLUE
TAILS	BLUE	RED

Table 2 Green payoff

	0	1	2	3	4	5	6	7	8
0	−9	−9	−9	−9	−9	−9	−9	−9	−9
1	0	0	0	0	0	0	0	0	0
2	15	15	15	15	15	15	15	15	15
3	36	36	36	36	36	36	36	36	36
4	63	63	63	63	63	63	63	63	63
5	96	96	96	96	96	96	96	96	96
6	135	135	135	135	135	135	135	135	135
7	180	180	180	180	180	180	180	132	132
8	231	231	231	231	231	231	231	177	123

Table 3 Blue payoff

	0	1	2	3	4	5	6	7	8
0	−9	−9	−9	−9	−9	−9	−9	−9	−9
1	0	0	0	0	0	0	0	0	−12
2	15	15	15	15	15	15	15	−3	−21
3	36	36	36	36	36	36	12	−12	−36
4	63	63	63	63	63	33	3	−27	−57
5	96	96	96	96	60	24	−12	−48	−84
6	135	135	135	93	51	9	−33	−75	−117
7	180	180	132	84	36	−12	−60	−108	−156
8	231	177	123	69	15	−39	−93	−147	−201

Table 4 Red payoff

	0	1	2	3	4	5	6	7	8
0	3	−3	−9	−15	−21	−27	−33	−39	−45
1	12	0	−12	−24	−36	−48	−60	−72	−84
2	15	−3	−21	−39	−57	−75	−93	−111	−129
3	12	−12	−36	−60	−84	−108	−132	−156	−156
4	3	−27	−57	−87	−117	−147	−177	−177	−177
5	−12	−48	−84	−120	−156	−192	−192	−192	−192
6	−33	−75	−117	−159	−201	−201	−201	−201	−201
7	−60	−108	−156	−204	−204	−204	−204	−204	−204
8	−93	−147	−201	−201	−201	−201	−201	−201	−201

The Experimental Design

Economic Decision-Making (EDM)

The EDM Experiment was held on Friday, February 27th 2015 at SBM Building – Bandung, Indonesia. Originally, the experiment was designed for three sessions. Each session contained of 20 iterations and two-set of participants (each set consist of 6 participants). However, due to extraneous obstacles, the experiment only executed for two sessions with total 24 respondents. All of the participants were postgraduate students. Specifically, the morning sessions consisted of MBA students, while in the afternoon session, the participants were more varied (MBA students and MSM[1] students).

Before entering the experimental situation, all participants were briefly explained about what the experiment about, how the experiment will be conducted and rules of the experiment.

Instructions for EDM Experiment

This is a Laboratory Economic Decision-Making. If you make good decisions, you might earn a CONSIDERABLE AMOUNT OF MONEY, which will be PAID TO YOU IN CASH at the end of the experiment.

There are two tasks in a sequence of periods: (i) the first period will be for practice, and (ii) the real remaining periods will be conducted for REAL MONEY that you will keep.

The experiment will be run for 5 periods with 5 rounds in each period. For each round, a person will meet 5 persons in dyadic. First task for each period is to state your goal in term of payoff (how much money that you will expect to achieve in this period?). Please state between negative 204 Rupiah (maximum lost) and 231 Rupiah (maximum gain).

For each period and for each round, choose whether or not you want to share some information with the person you are paired with (in term of toss coin). For each period and each round, make choice from the integers 0 to 8. Your choices and the choices of the person you are paired with will determine the payoff you earn for that round. Moreover, the payoff you earn for that period is the average of each round.

There are three possible "payoff tables": (i) GREEN table (25 % chance), (ii) BLUE (50 % chance), and (iii) RED (25 % chance)

Experiment of Seating Design

Each session of the experiment consists of 12 participants. Every participant cannot see their opponent or their pairing for certain round (see Fig. 1).

Every participant is arranged in certain manner that in every round they will pair with different participants. For every set (every six participants), the pairing will be:

---

[1] MSM = Master of Science in Management

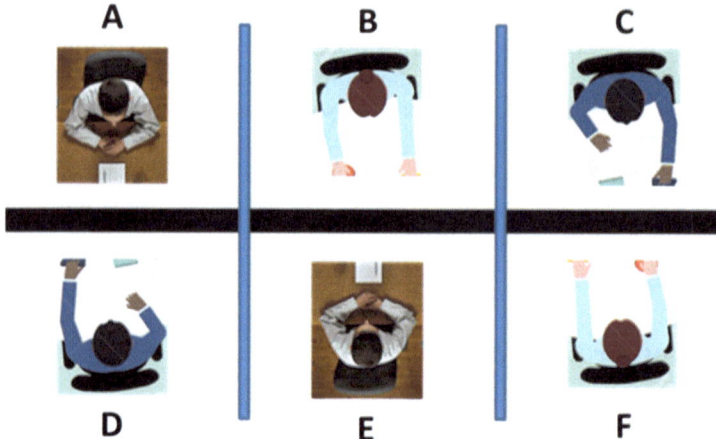

Fig. 1 Seating arrangement for every set of experiment

Table 5 Pairing participants

A with B	A with C	A with D	A with E	A with F
B with C	B with D	B with E	B with F	C with D
C with E	C with F	D with E	D with F	E with F

Round 1

D	B	C
E	F	A

Round 2

D	B	C
F	A	E

Round 3

A	B	C
E	D	F

Round 4

A	B	C
F	E	D

Round 5

B	F	A
C	E	D

Fig. 2 Seating arrangement for each iteration

Based on the combination on Table 5, the seating arrangement can be described as Fig. 2.

In the beginning of each iteration, the subject will be asked by the target he/she wants to achieve in this iteration and it is recorded in the paper as in Table 6.

After that, the subject flips the coin and decides whether they want to share (S) or not to share (NS) and record the sharing information in the paper as in Table 7.

The subject will give Table 7 sheet to the instructor and based on this, the instructor will determine which table will be used and whether the subject will get the table (depend on the decision from both subjects). The subject will decide the

Table 6 Aspiration filled sheet

#Iteration	Session	
Aspiration:		

Table 7 Sharing information filled sheet

Session		Iteration		Round
H/T		S/NS		1

Table 8 Production filled sheet

Session				
#Iteration				
R1	R2	R3	R4	R5

amount of quantity they want to produce in a certain round (R) and record it on the sheet as described at Table 8.

After each round, the subject will be arranged according to the formation on Fig. 2. In the end of each iteration, the subject will receive their aggregate payoff.

Experiment Result

Behavioral Result

During the experiment, the participants were told to flip a coin and the result was affected on the type of Payoff Table. Based on the two sessions and five iterations in each session, the recapitulation of the payoff table can be shown on Table 9 (a = morning session; b = afternoon session).

Based on the Table 2, the behavioral between respondents in two sessions was compared. The hypotheses tested were:

1. *Ha: There is difference between the proportion of GREEN payoff-table between morning and afternoon session*
2. *Ha: There is difference between the proportion of BLUE payoff-table between morning and afternoon session*
3. *Ha: There is difference between the proportion of RED payoff-table between morning and afternoon session*

The result for each hypothesis as follows.

1. Z-stat is 2.344895 whereas the *Z-table* is 1.96 ($\alpha = 0.05$). So based on the statistical test, there is a significant difference between the proportion of GREEN payoff-table between morning session and afternoon session.
2. *Z-stat* is -0.58109 whereas the *Z-table* is -1.96 ($\alpha = 0.05$). Based on the statistical test, there is no difference between the proportion of BLUE payoff-table between morning session and afternoon session.

Table 9 Payoff table occur during experiment

(a)

Payoff table	Iteration				
	1	2	3	4	5
Green	16	18	12	14	12
Blue	13	10	14	14	13
Red	1	2	4	2	5
	30	30	30	30	30

(b)

Payoff table	Iteration				
	1	2	3	4	5
Green	12	11	7	13	9
Blue	14	15	16	9	15
Red	4	4	7	8	6
	30	30	30	30	30

Table 10 Aggregate payoff in each session

Session	Iteration					Average payoff
	1	2	3	4	5	
Morning	77.65	91.75	48.25	63.85	67.4	69.78
Afternoon	44.35	68.92	35.45	28.7	57.3	46.94

3. *Z-stat* is -2.47145 whereas the *Z-table* is -1.96 ($\alpha = 0.05$). Based on the statistical test, there is a significant difference between the proportion of RED payoff-table between morning session and afternoon session.

According to the hypothesis test result, suspected that the respondents in the morning session tend to not comply with the rule of coin flip (e.g. they just circle the sharing information sheet – Head or Tail option, instead of flipping the coin first then filling the sharing information sheet) after they recognize that the red payoff table provided is loss. The aggregate payoff between sessions is shown at Table 10.

In morning session, the average payoff (69.78) is higher than symmetric static Nash equilibrium on blue payoff table (63). Although the average of green payoff (14.4) is greater than the average of blue payoff (12.8), the subjects cannot achieve the average payoff close to symmetric static Nash equilibrium on green payoff table (180). This situation shows that the subjects are not willing to share their flipped coin. If they share the flipped coin, they know that they use the green table and try to coordinate their action. On the other hand, the afternoon session shows the average payoff (46.94) is lower than symmetric static Nash equilibrium on blue payoff table (63). The average of green payoff (10.4) is lower than the average of blue payoff (13.8). This result also shows the lack of coordination (does not share the information). From these results, it is showed that the collusive behavior does not emerge if the subjects do not have the fixed partner.

Table 11 Grand mean of aspiration

Session	Iteration					Grand mean
	1	2	3	4	5	
Morning	176.56	160.48	169.42	147.28	178.69	166.49
Afternoon	128.29	83.47	98.78	83.75	53.75	89.61

Aspiration Level

Table 11 shows the average and grand average of aspiration level of the subject from both sessions. As we can see, the average of the morning session is higher than the afternoon session. The subjects in the morning session are not more sensitive to change their aspiration based on their last aspiration. It can be shown in Table 12a, b). The subjects in the afternoon session are very sensitive in determining their aspiration based on their previous aspiration. They set their aspiration high, if they get high payoff in previous iteration and vice versa. The subjects in the afternoon session also tend to have competitive environment.

Willingness to Share

In this session, we want to investigate the relationship between the result of flipped coin and willingness to share. The result on the EDM on participant willingness sharing the result of the coin-flipped is shown in Table 13:

Based on the result, the χ^2_{stat} for morning session is 0.045 and χ^2_{stat} for afternoon session is 0.217, whereas the χ^2_{table} is 3.41. It means that there is no relationship between the result of flipped coin and willingness to share the information both in the morning and afternoon sessions. From these results, the subjects tend to ignore the chance to get better payoff. On the other hand, if the subject is aware of the result of the coin-flipped and with this awareness they will share the result, the subject will reduce the uncertainty. The uncertainty is about what kind of outcomes they will get, which depends on the payoff table.

Table 12 Aspiration and payoff of the subjects in (a) morning session and (b) afternoon session

(a)

Respondent	Iteration 1		Iteration 2		Iteration 3		Iteration 4		Iteration 5	
	Aspiration	Payoff	Aspiration	Payoff	Aspiration	Payoff	Aspiration	Payoff	Aspiration	Payoff
A1	115.5	43.8	200.8	93.6	225.6	12.6	225.6	56.4	220.5	57.6
B1	220	81.6	231	134.4	231	-72.6	231	30	230	9.6
C1	75	107.4	80	82.2	85	82.8	100	79.8	80	57
D1	200	96.6	230	36.6	210	27	180	52.8	200	103.2
E1	222.2	52	130	47	140.4	-41	22.2	22	210.3	28
F1	135	48.6	231	133	135	-24	0	9.6	210	67.8
A2	200	43.8	180	97.8	220	53.4	190	61.2	180	63
B2	200	34.8	55	69.6	55	40.2	50	93	100	48.6
C2	100	39.6	123	119	125	182	126.5	84	126.5	78.6
D2	220	125	230	82.8	225	96.6	231	100	231	99
E2	231	92.4	135	134	231	97.2	231	53.4	231	109
F2	200	167	100	70.8	150	125	180	124	125	88.2

(b)

Respondent	Iteration 1		Iteration 2		Iteration 3		Iteration 4		Iteration 5	
	Aspiration	Payoff	Aspiration	Payoff	Aspiration	Payoff	Aspiration	Payoff	Aspiration	Payoff
A1	150	58.2	80	46.2	80	18	80	63	70	24
B1	200	12	100	85.8	100	18	100	-54	200	142
C1	210	12	25	85.8	52	1.8	30	-54	50	142
D1	100	-26	80	39.6	80	46.2	80	-17	80	68.4
E1	30	10.8	50	118	70	-6.6	70	-15	100	33.6
F1	100	-26	185	39.6	129	46.2	30	-17	45	68.4
A2	207.5	109	125	48.6	100	135	125	33.6	100	102
B2	70	81.6	75	128	100	40.8	80	37.2	90	78.6

C2	170	107	123	66	77	31.8	250	-4.2	38	99.6
D2	202	15.6	31	77.4	220	-26	3	56.4	-24	91.8
E2	100	18	100	50.6	100	83.4	70	123	100	-10
F2	0	27.6	27.6	77.4	77.4	87	87	56.4	-204	-10

Table 13 Summary of willingness to share based on the result of flipped coin in (a) morning session and (b) afternoon session

(a)				(b)			
	Decision				Decision		
Flip coin result	Share	Not share	Total	Flip coin result	Share	Not share	Total
Head	148	46	194	Head	127	46	173
Tail	72	21	93	Tail	100	32	132
Total	220	67	287	Total	227	78	305

Conclusion

The experimental results show that the effects of sharing information bring the market close to symmetric static Nash equilibrium. However, the subjects tend to ignore their past experience to improve their outcome. It means that the subjects do not learn through the experiment. The subjects also ignore the importance of sharing the result of flipped coin. They do not know that the payoff table will reduce the uncertainty about the outcome, if they share the result of flipped coin. This behavior does not change through the experimentation; it means that they do not learn. In term of aspiration, the experiment shows that the subject tends to set their aspiration level based on what they achieve in previous iteration. However, they set their aspiration with extremely bigger value than other iterations.

References

Altavilla, C., Luini, L., & Sbriglia, P. (2006). Social learning in market games. *Journal of Economic Behavior and Organization, 61*, 632–652.

Bensen, B. L., & Faminow, M. L. (1988). The impact of experience on prices and profits in experimental duopoly markets. *Journal of Economic Behavior and Organization, 9*, 345–365.

Cason, T. N., & Davis, D. D. (1995). Price communications in laboratory markets: An experimental investigation. *Review of Industrial Organization, 10*, 769–787.

Cason, T. N., & Mason, C. F. (1999). Uncertainty, information sharing and tacit collusion in laboratory duopoly markets. *Economy Inquiry, 37*, 258–281.

Davis, D. D., & Holt, C. A. (1993). *Experimental economics*. Princeton: Princeton University Press.

Dixon, H. (2000). Keeping up with the joneses: Competition and the evolution of collusion. *Journal of Economic Behavior and Organization, 43*, 223–238.

Dixon, H., Sbriglia, P., & Somma, E. (2006). Learning to collude: An exper- iment in convergence and equilibrium selection in oligopoly. *Research in Economic, 60*, 155–167.

Hardstad, R., Martin, S., & Normann, H.-T. (1998). Intertemporal pricing schemes. Experimental tests of consciously parallel behavior in oligopoly. In L. Philips (Ed.), *Applied industrial economics*. Cambridge: Cambridge University Press.

Holt, C. A. (1985). An experimental test of the consistent-conjectures hy- pothesis. *The American Economic Review, 75*(3), 314–325.

Huck, S., Muller, W., & Normann, H. T. (2001). Stackelberg beats Cournot -on collusion and efficiency in experimental markets. *The Economic Journal, 111*(474), 749–765.

Huck, S., Normann, H., & Oechssler, J. (1999). Learning in cournot oligopoly an experiment. *Economic Journal, 109*, C80–C95.

Lupi, P., & Sbriglia, P. (2003). Exploring human behavior and learning in experimental cournot settings. *Rivista Internazionale di Scienze Sociali, 111*, 373–395.

Simon, H. (1955). A behavioural model of rational choice. *Quarterly Journal of Economics, 69*, 99–118.

Simon, H. A. (1956). Rational choice and the structure of the environment. *Psychological Review, 63*(2), 129–138.

Vega-Redondo, F. (1997). The evolution of walrasian behaviour. *Econometrica, 65*(2), 375–384.

Value Co-creation on Cloud Computing:
A Case Study of Bandung City, Indonesia

Santi Novani

Abstract Cloud computing is a type of service system in today's IT world. There is a critic to Good Dominant Logic, which defines cloud service as a single, uni-directional activity during the operating phase. The aim of this research is to describe value co-creation based on Service Dominant Logic in how to deliver services to the society in Bandung city, Indonesia. Most governments only focus on internal operations which support government activities rather than on optimizing service delivery based on society's needs. Therefore, in this research, we will emphasize on how cloud can support Bandung city based on value co-creation processes, which involve the customers demand (society) to deliver improving citizen services.

Introduction

Service is defined as the application of benefit to create value, which is derived from the interaction among entities as a service system. A service system involves four components, i.e., people, technology, organizations and shared information (Spohrer and Maglio 2009). In the global era, service systems and service networks use digital connection based on information and communication technologies (ICT) to share and accelerate values from knowledge (Demirkan and Delen 2013). When service and ICT is discussed, the concept of cloud computing should be introduced (Armbrust et al. 2009; Buyya et al. 2008; Li et al. 2011; Shen et al. 2011), which is a type of service system in today's IT world (Demirkan and Delen 2013).

Cloud computing is a service which provides resources, software and information as a utility over network to end-users. Meanwhile, National Institute of Standards defines cloud computing as a model to enable convenient, network access to shared configurable computing resources with minimum service provider interactions to release rapidly (Demirkan and Delen 2013). Cloud computing is an IT platform which provides an enabling technology and provision of services in service oriented environment (Haile and Altzman 2012). There are three models

S. Novani (✉)
School of business and Management, ITB, West Java, Indonesia
e-mail: snovani@sbm-itb.ac.id

© Springer Japan 2016

43

K. Mangkusubroto et al. (eds.), *Systems Science for Complex Policy Making*,
Translational Systems Sciences 3, DOI 10.1007/978-4-431-55273-4_4

of services in cloud computing (Clouds), i.e., Infrastructure as a Service (IaaS), Platform as a Service (PaaS) and Software as a Service (SaaS).

There is a critic to G-D logic which defines cloud service as a single, uni-directional activity during the operating phase of cloud service. According to Brynjolfsson et al. 2010, cloud computing does not fit to the utility model and that there is the co-creation of value in cloud computing. However, to our best knowledge, there are no mechanisms for value co-creation in cloud-computing that have been described. Some of cloud computing will become the founding infrastructure of economies. Therefore, this paper introduces a SD-Logic (Lusch and Vargo 2006) based approach for value co-creation in cloud computing since it is important to understand it.

In developed country such as USA, Government organizations are redefining their businesses to deliver improved citizen services. According to the US Federal Cloud Computing Strategy, the US Government instituted the Cloud first policy to accelerate the pace of cloud adoption. In this research, Indonesia as a developing country will be discussed as how to implement cloud computing. Indeed, cloud computing has been popular in Indonesia since 2010 and it is relatively new. The question is why do we need a cloud? There is no doubt that customers want it. Indonesia as a developing country would be very suitable for the development of the cloud computing business, with cost-efficiency reasons.

At the present time, about half of an IT department's budget goes to traditional IT, according to IDC, with 16 % being outsourced and the remaining 35 % going to public and private clouds in Indonesia (IDC 2010). Two years from 2010, the percentages changed significantly. Traditional IT drops to 37 %, outsourced IT remains the same, yet the cloud jumps to 47 %.

Cloud was introduced by PT Telekomunikasi Indonesia (PT Telkom) about 2006 through its subsidiary TelkomSigma (Mangula et al. 2012). The development of cloud has got support by the establishment of Indonesian Cloud Forum (ICF) in 2011. ICF is a community forum of users, practitioners and providers, which focus on cloud issues in Indonesia. The trend on cloud adoption is increasing from years to years in Indonesia. In the future, it is estimated that more companies and consumers in Indonesia will use cloud computing services.

The importance of cloud computing is expressed by companies, and the emergence of cloud computing in the country indicates a steady progress. Many companies in Indonesia working with big data are starting to turn on cloud computing, especially those small companies, which do not have the resource to buy, maintain and secure their information systems. Cloud computing services will be the solution to the problems.

The aim of this research is to describe value co-creation on cloud computing to deliver services to the society in Bandung city, Indonesia. Bandung society wants a better accountability, transparency and service delivery from city governments. Most government only focuses on internal operation which supports government activities rather than on optimizing service delivery based on society's needs. Therefore in this research, we will emphasize on how cloud can support Bandung city based on value co-creation process which involve the customers demand (society) to deliver improved citizen services.

Indonesia Cloud Computing

In this part, we discuss the existing cloud computing profile in Indonesia, which consists of cloud readiness, Indonesia Internet users and then followed by the information about cloud provider in Indonesia.

Cloud Readiness in Indonesia

Cloud computing has become a global trend in computing in the last few years. Cloud computing emerged in Indonesia since 2008, this phenomenon is interesting and still being defined and open to be interpreted. According to Haag and Cummings (2010), cloud computing is a technology model of which any and all resources-application software, processing power, data storage, backup facilities, development tools and literally everything are delivered as a set of services via Internet connection.

Based on the survey conducted by Springboard Research (2011), the use of cloud computing in Indonesia is only 20 %. If we compared it with other countries such as Singapore 29 %, Malaysia 27 % and Korea 55 %, this means that there is still a great potential and opportunity to develop cloud computing in Indonesia. Indonesia as a developing country would be very suitable for the development of the cloud computing business, for cost-efficiency reasons. There are some advantages of cloud computing, such as economies of scale resulting in low-costs of IT infrastructure, improving the performance to access more dynamic, offering easier data monitoring, low cost to undertake security measures and memory and storage capabilities based on the demand (Voona and Venkantaratna 2009; Buyya et al. 2008; Miller 2008; Catteddu and Hogben 2009; Andrei 2009).

However cloud also has a number of disadvantages such as requiring a constant internet connection, limited features offered, security might not meet the organization standards and danger of loss of business in case of data loss (Miller 2008; Jeffrey and Neidecker-Lutz 2009; Ristenpart et al. 2009).

In Indonesia, the market of cloud computing grew by 43 % in 2012 with revenue $31.4. Most of business in Indonesia have been using cloud computing. According to the cloud readiness index (2014) which is described in Fig. 1, Indonesia was ranked as the 12th with CRI score 52.4. Cloud computing is seen as an effective way to enhance the efficiency of IT infrastructure.

In the future, it is estimated that more companies and consumers in Indonesia will use cloud computing services, especially, with the increasing number of smart phones and tablet computers users. In Indonesia, cloud computing market is

Cloud Readiness Index 2014

	1. Privacy	2. International Connectivity	3. Data Sovereignty	4. Broadband Quality	5. Government Regulatory Environment and Usage	6. Power Grid and Green Policy	7. Intellectual Property Protection	8. Business Sophistication	9. Data Centre Risk	10. Freedom of Information	CRI2014 SCORE	RANK	CHANGE
JP Japan	9.5	5.5	8.0	9.1	5.0	7.1	8.1	8.2	6.6	9.7	76.8	1	-
NZ New Zealand	8.8	4.6	7.9	7.6	5.6	9.2	8.6	6.8	7.8	9.5	76.3	2	+4
AU Australia	8.8	4.4	7.6	8.0	5.3	7.8	7.6	6.7	9.4	9.6	75.1	3	+4
SG Singapore	6.0	8.2	7.8	8.8	6.1	5.9	8.7	7.3	7.4	8.6	74.8	4	-
HK Hong Kong	6.8	7.7	7.6	9.3	5.1	5.6	8.1	7.5	7.4	9.6	74.7	5	-2
KR South Korea	9.7	5.5	7.2	9.4	5.1	6.6	5.7	6.9	8.6	8.6	73.3	6	-4
TW Taiwan	4.6	6.3	6.8	8.5	5.0	6.7	7.4	7.4	6.9	8.6	68.2	7	-2
MY Malaysia	5.8	5.8	6.7	7.1	5.2	4.9	6.9	7.2	8.5	8.2	66.2	8	-
TH Thailand	4.0	5.0	6.2	8.0	3.7	6.3	4.4	6.3	7.6	7.8	59.3	9	+4
PH Philippines	5.8	5.4	5.9	4.1	3.7	5.5	5.1	6.1	5.5	9.0	56.1	10	+2
CN China	5.9	3.0	4.8	5.9	4.3	4.3	5.6	6.2	6.5	7.0	53.3	11	-1
ID Indonesia	4.4	2.9	6.2	3.1	3.9	5.7	5.6	6.3	6.4	7.9	52.4	12	-1
IN India	4.6	2.3	6.5	3.6	4.1	5.0	5.3	6.3	3.4	7.8	48.8	13	-4
VN Vietnam	3.6	3.2	5.6	4.2	3.8	4.7	4.1	5.3	6.4	7.0	47.8	14	-1

Source: Asia Cloud Computing Association 2014

Fig. 1 Cloud readiness index (2014) taken from Asia Cloud Computing Association

expected to reach revenue in excess of $150 million by 2016. The major verticals that are set to boost the growth include manufacturing, telecommunication and service providers, content and media businesses, government and education. In a few short years, cloud computing has become an important part of the knowledge economy and certainly become one of the biggest drivers of economic growth over the next decade.

Meanwhile, the readiness of Indonesia, for some assessment criteria, it is still below the average value. These criteria include the data sovereignty, broadband quality, online government service and ICT prioritization, power grid and green policy and data center risk. To see further readiness of cloud computing services in Indonesia, some conditions related to ICT services in Indonesia will be elaborated below.

Indonesia's Internet User

Until 2012, Internet users in Indonesia tend to increase. Individual Internet users and more and more enterprises in Indonesia are jumping on the bandwagon. The emergence of cloud computing in the country also indicates a steady progress. According to Frost and Sullivan (Jakarta post 2014), this market will experience significant growth even though nowadays the use of cloud computing in the country is still considered low. As quoted by Frost and Sullivan (Jakarta Post 2014), the

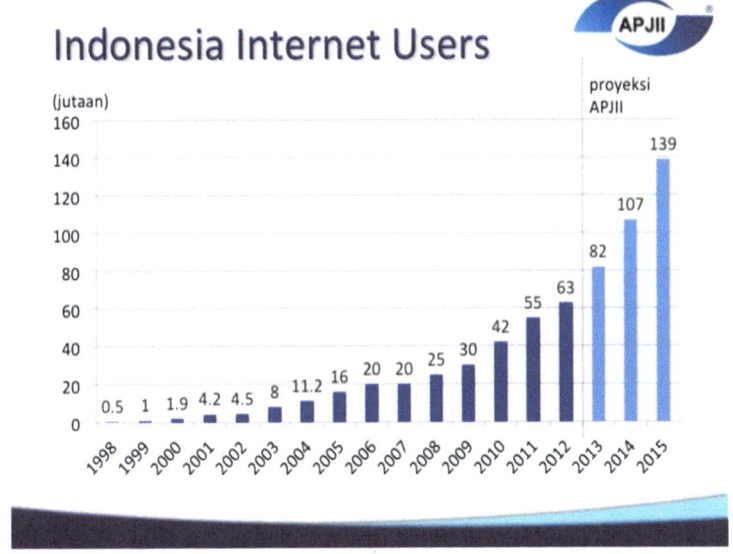

Fig. 2 Indonesia Internet users (Source: The Association of Indonesian Internet Providers (APJII), 2013)

market growth of cloud computing in Indonesia would reach more than $120 million by 2017 and that Software-as-a-Service (SaaS) would be the popular choice of Indonesians.

Figure 2 shows the development trend of Internet users in Indonesia based on data from APJII (Association of Indonesian Internet Service Providers).

From 2013 to 2015, based on APJII projections, Indonesian Internet users increased from 82 million to 139 million people. This rapid growth also shows the shift in the mindset of Indonesian people, especially business people. Many young entrepreneurs are successful through online presence. Marked by the appearance of digital startup in Indonesia, SMEs are increasing internet literate of the students who are going into business via the internet.

Cloud Computing Provider in Indonesia

Trends in the development of the use of Cloud Computing Services in Indonesia will be in parallel with the development trend in other countries. Cloud computing business in Indonesia is widely open. Providers are competing to get the profit in these new businesses. Cloud providers try to use their own methods to meet the expectations from the organization or business people who need their services. There are several cloud computing providers who are actively competing in offering their services to meet the organization needs, such as TelkomSigma, Indosat,

Table 1 Indonesia cloud service

Provider	SaaS/App	PaaS	IaaS/Cloud server	DraaS	Cloud storage
TelkomSigma	Yes	No	Yes	Yes	No
Indosat	No	No	Yes	No	No
XCloud	No	No	Yes	No	Yes
Bizznet	No	No	Yes	No	Yes
Indonesia cloud	Yes	No	Yes	Yes	Yes

Biznet, XL, Indonesian Cloud, PT Infinys System Indonesia, Huawei Cloud, HP Cloud Computing Solution, etc. Cloud computing services in Indonesia can be categorized as Iaas (Infrastructure as a service), SaaS (Software as a Service), PaaS (Platform as a Service), DraaS (Disaster Recovery as a Service), Cloud Storage and other services.

Table 1 shows some providers with their cloud computing services.

Cloud Computing: Service System Science Perspective

Cloud computing is a new way of delivering computing services. In the previous study (Dachyar and Prasetya 2012), cloud is defined as a new technology. In this study, it is also already discussed about what factors are associated with significant recommendations of use of new technologies by using quantitative research. Since it is still new, it opens the opportunity to provide a good service to customers. Cloud computing is an emerging paradigm shift in management of computing resources for provisioning of services with enhanced capabilities.

In this study, by using the service system science perspective, cloud should maintain and affect relationships among stakeholders in specific industry contexts, as well as identify how the stakeholders will interact by using value co-creation process and how their roles may shift, where the economic value proposition produced should be unique.

Cloud Computing as a Service System

In this part, we discuss cloud as a service system which is defined as dynamic interactions of providers, customers, ICT (information and communication technology) and share information that creates value between the provider and the customer (Cambridge White Paper 2007). In organizations, cloud computing is a good tool to improve the state of other systems and it acquires external resources by integrating client resources or by receiving and returning them back after the usage (Maglio et al. 2006). It means that cloud is a good tool to interact and collaborate with customers (Edlund 2012), or in other words, it offers effective tools for co-creation to its customers.

Organizations have come to realize that cloud is a good way to interact and collaborate, but how to realize it is still hotly to be discussed. In this study, we use the concept of service system to identify the interaction among the stakeholder in cloud computing to create the value. Value co-creation is the heart of S-D logic, which emphasizes that we need active participation from the customers along with the provider to create the value in the marketplace.

Some questions are raised about how to engage the customer to participate in and how and where co-creation actually takes place. Many previous researches on example of co-creation by using an Internet, such as electronic services (Blazevic and Lievens 2008) and virtual co-creation, co-creation of virtual communities (Zwass 2010), the raise of social media and Internet, enhanced the opportunity for provider and customers to collaborate (Sawhney et al. 2005; Deighton and Kornfeld 2009; Hennig-Thurau et al. 2010; Novani and Kijima 2012).

According to Gronroos (2008), the firm is fundamentally a value facilitator, which means it implies the firm's responsibility to provide customer with all resource for value co-creation. This argument is also supported by Norman and Ramirez (1993) that the key to creating value is to mobilize customers. Therefore, providing customers with platforms which are convenient and easy to use would significantly enhance the opportunity to a fruitful co-creation. To access and share these resources, a firm and a customer need an "encounter platform" (Yazdani 2012). In this study, we define that cloud service can be used as an encounter platform to provide value co-creation process.

According to Willcocks et al. (2011), customers experience in cloud computing is important. Cloud will increase the importance to deliver the service and the quality of customers' experience. Nowadays, customers are active and they have a high expectation to, therefore cloud must raise their level of service quality, security and competence as well.

Cloud is a way to engage in co-creation process and it significantly enhances customers experience throughout the collaboration process. Cloud should be considered a tool to facilitate and manage collaboration as a core process of value co-creation.

Cloud Value Co-creation

This part explains the co-creation process, which means the role of customers is important in order to interact with provider. According to the premised of S-D logic, the value is not pre-determined but defined and co-created by customers and provider through interaction (Prahalad and Ramaswamy 2004; Vargo and Lusch 2004). In S-D logic, customers are active participants in the production and delivery of cloud computing service and engaging in the process of value co-creation (Ostrom et al. 2010; Vargo and Lusch 2004). In the traditional computing, the

 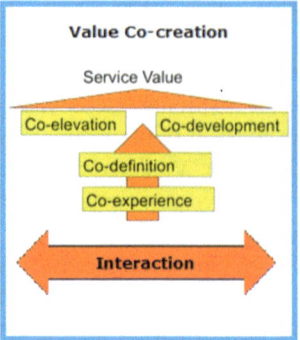 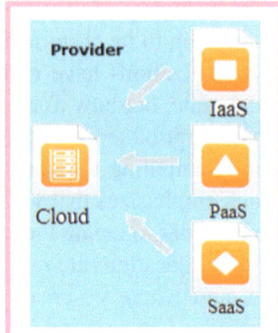

Fig. 3 Cloud value Co-creation

resources which are delivered to customers is tangible goods, meanwhile in cloud computing which is delivered in digital form is a service (Fig. 3).

In the value co-creation process, it consists of four stages, i.e., co-experience, co-definition, co-elevation and co-development (Galbrun and Kijima 2009). We briefly define and describe each underlying concept used to describe the service innovation process.

1. *Co-experience*
   The intent of experience innovation is not to improve a product or service, per se, but to enable the co-creation of an environment in which personalized and evolvable experiences are the goal, and products and services are a means to that end.
2. *Co-definition*
   Despite the potential value of user-generated innovation, it can be difficult for firms to access and integrate user-based knowledge. User based knowledge is tacit, based on experience and thus difficult to transfer (Von Hippel 1988). Consequently, "spill-over" of such knowledge and their artifacts, patents and licenses are therefore limited to allow knowledge transfer between customer and provider. We argue that this collaboration with users requires a co-definition process by which a shared knowledge model is built by provider, lead customer, where learning from each other is pivotal (Metcalfe et al. 2005; Lusch et al. 2009). Satisfaction is generated by co-experiencing and co-defining a shared internal model by provider and customer.
3. *Co-elevation*
   In the effort to further develop the service innovation process, it is relevant to relate it explicitly to general systems theory (Bertalanffy 1968), which has been used much more in the natural sciences than in the social sciences. In generic terms, by system we mean that a set of two kinds of constituents. On the one hand, the components or entities of the system, and on the other hand, the relations among them form a coherent whole. The former, we call it co-elevation, focuses more on co-innovation led by some particular entities

among the system. Co-elevation is first described as spiral up elevation of the expectation from customer or abilities of provider.

4. *Co-development*

Alternatively, the later, namely co-development drives our attention to co-innovation generated by the relations among the various entities. The cognitive gap among entities is the driver of a collaborative process where entities exchange heterogeneous knowledge bases to contribute to a co-development of solutions. In this case, the intensity and the variety of interactions matters described by some scholars as network dynamics or value network (Christensen and Rosenbloom 1995; Lusch et al. 2009; Prahalad and Ramaswamy 2004).

Cloud Value Co-creation in Bandung City

In this part, we will discuss the ICT condition in Bandung. The assessment of its existing condition is required to study cloud computing which will be applied to the public sector and communities.

Vision and Mission of Bandung City

ICT vision of Bandung is *Towards Interactive e-Government and Leading Bandung as a Smart City* which in turn is based on the Medium- and Long-Term Development Platform ICT sector. This vision is expected to be achieved by 2025, when people will be able to easily access the full-quality public services, safely and comfortably through any e-media, anywhere and anytime. In accordance with the stage of development of ICT, in 2018, Bandung is expected to reach the stage of interactive ICT. Therefore, ICT vision of Bandung in 2018 is *Towards Interactive e-Government and Leading Bandung as a Smart City*.

ICT vision of Bandung becomes the basis to identify the mission, direction and determination of the ICT requirement of Bandung. ICT vision of Bandung contains three main elements:

1. Interactive e-Government

This element describes the condition of interactive e-government, both between SKPDs/agencies within city government (G to G) and between the government and the business/community (G to B and G to C), with assurance on security and privacy.

2. Leading Bandung

Bandung Juara (Bandung Champion) is a vision of development from 2014 to 2018. *Bandung Juara* means a winner, powerful and prosperous. In this case, the e-government aims to support the achievement of the development according to the vision and mission development.

3. Bandung Smart City
   These elements describe ICT as a major element in realizing the concept of smart city development initiatives with a focus on executive dashboards, paperless office, as well as public services online.

ICT Needs Identification of Bandung

ICT Bandung needs are based on the vision and mission of ICT Bandung, which is divided into two parts:

1. Management of Internal Governance
   ICT needs associated with internal management governance encompass the following features:

   (a) Provision of information processing facilities
   (b) Provision of monitoring facilities for city leaders
   (c) Telecommunications and collaboration facilities on SKPDs

2. *Public Service*
   ICT needs for public services encompass the following:

   (a) ICT facilities for quality and affordable public services
   (b) Facilities for the provision of public services particularly information for the community
   (c) ICT facilities to improve the quality of social and economic

Based on the identification of ICT needs of Bandung, ICT development is focused on the following five points:

1. Infrastructure policy, so that the Internet penetration becomes increasingly faster, easier and affordable.
2. Facility to monitor information for the senior leaders of the city and the community, in the form of dashboards and portals.
3. Route ware net (RWNET) development as a center for information exchange community.
4. The development of educational content and creative industries.
5. Provision of facilities for the cable network (ducting) to ease the development of infrastructure.

To realize all points, the ICT development is mapped into ten functions in accordance with the directives of Communication and Information Biro/Diskominfo Governance. The ten functions are divided into two groups, namely the functions that are the core of e-government and the function of Local Government Task Force (*Satuan Kerja Perangkat Daerah* – SKPD) (Fig. 4. Functional SKPDs Information Systems).

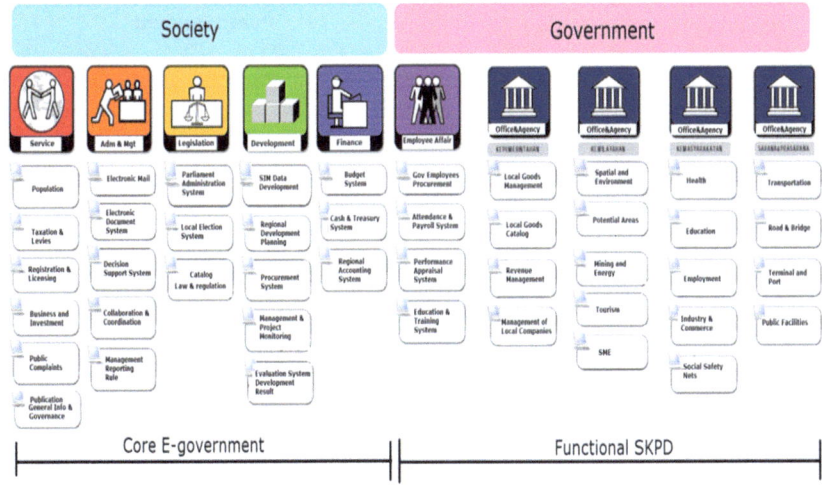

Fig. 4 Functional SKPDs information systems (Source: Bappeda Kota Bandung 2013)

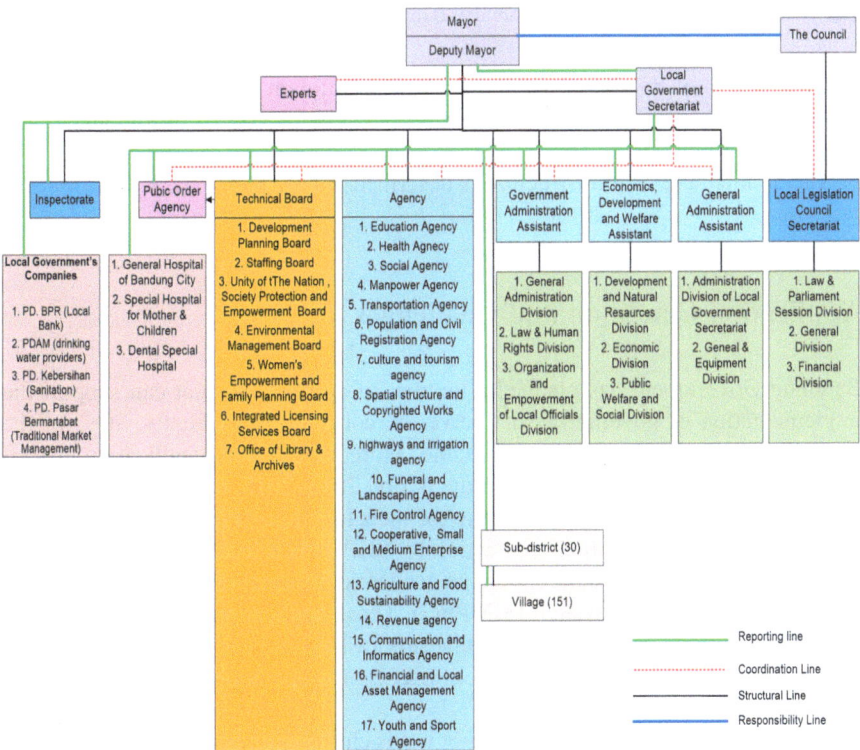

Fig. 5 Structure organization of Bandung Government (Source: www.bandung.go.id)

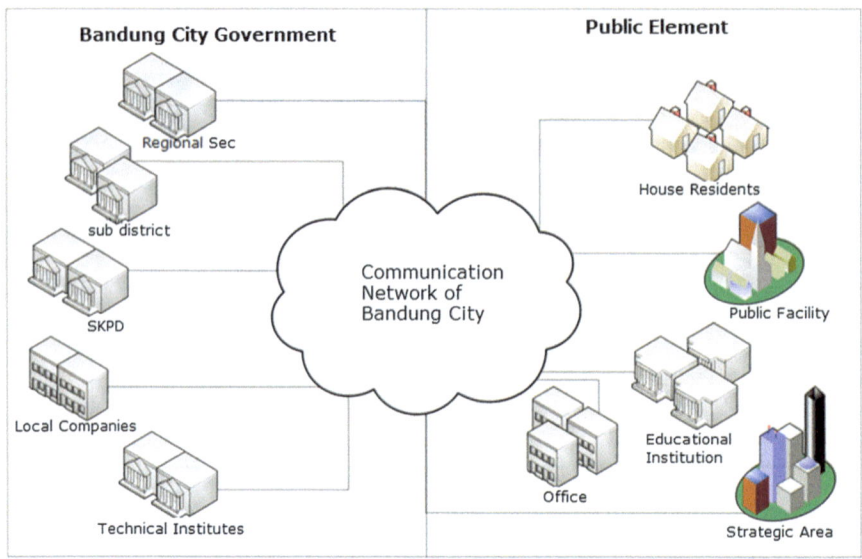

Fig. 6 Architecture of infrastructure (Source: Bappeda Kota Bandung 2013)

Government of Bandung has 78 SKPDs with different roles and responsibilities in running the government and the public service. Figure 5 shows the organizational structure of Bandung.

The architecture of infrastructure is described in Fig. 6.

Application Condition and Challenges

Bandung Government has planned a number of applications that can support the implementation of governance and serve the community. Currently, 70 % of the applications required are not yet available. Moreover, existing application utilization is not maximized. Therefore, in the period 2013–2018, development applications are prioritized due to its criticality of their functions (Fig. 7).

The most widely installed application by SKPDs is a local management information system (SIMDA) application, an academic information system (SIAK) and a personnel information system (SIMPEG). These applications support the activities of government. Application to public service has not been installed by SKPDs. The number of applications installed in SKPDs for operational activities are 1–5 applications at most.

Decision to adopt cloud computing or not is challenging because of the range of sociopolitical and practical reasons. The challenges that Bandung must address before cloud computing are providing the accurate information on costs of cloud adoption, supporting risk management and ensuring the decision makers can make informed tradeoff between benefit and risk (Khajeh-Hosseini et al. 2010).

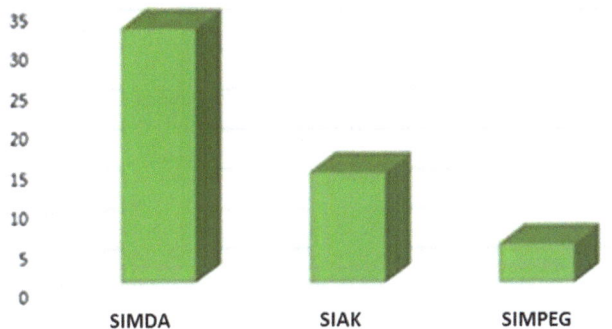

Fig. 7 Most widely installed applications (Source: Bappeda Kota Bandung 2013)

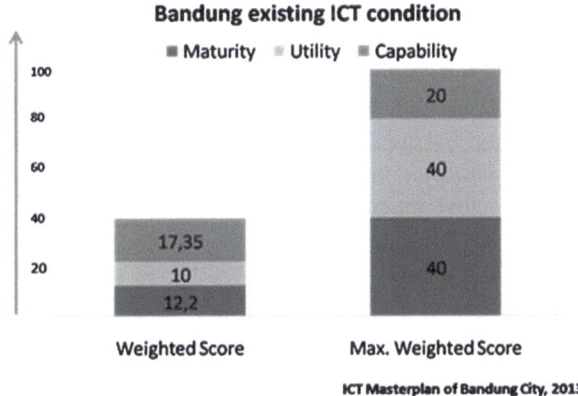

Fig. 8 Bandung existing ICT condition (2013)

The graph below shows the existing condition of ICT situation in Bandung. It is based on a survey (ITB) through all functional unit work in Bandung, regarding to the maturity, utility and capability significance of ICT towards Leading Bandung motive.

After maximum weighted score, the results depict that infrastructure and facility of ICT are not trade-off with the capability of human capital skills in Bandung and its functional work unit areas. One of the challenges is to maintain and train the human capital skills to ensure the proposed strategy runs well in Bandung afterward (Fig. 8).

In term of ICT implementation in the cities, Bandung is rewarded by 39 scores that indicates Bandung ranking in the transition level between enhance and interactive level. Although it is still above national benchmark score, Bandung can still improve the score by implementing new strategies and one of them is by using cloud computing strategy.

Related to Bandung government case as what has been described in the Leading Bandung Grand Design, cloud computing would be found as a useful, effective and

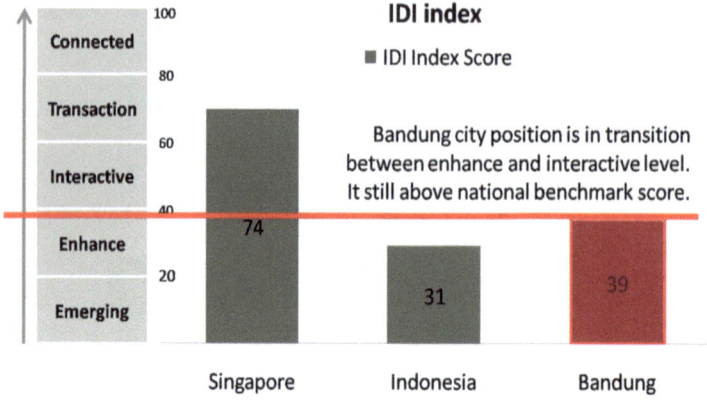

ICT Masterplan of Bandung City, 2013

Fig. 9 IDI index score

affordable way to improve both internal and external functions of service. Internally, Bandung can step by step increase the quality of the data administration, communication and monitoring in the heart of government or even in every work unit of Bandung City government to bring a better performance and decision-making process into reality.

Externally, Bandung can surely offer a better service to its citizens, including the potential business partners which indicate future needs of Bandung (Fig. 9).

Based on the road map of ICT implementation in ICT Master plan of Bandung, in the brand new visions and missions, it is showed that the city needs new IT solutions to improve all aspects in Bandung.

The Proposed Cloud Value Co-creation in Bandung

By exploring the existing condition of Bandung, it can be seen that Bandung needs new IT solutions to improve public service. In this study, we propose cloud as an IT solution. In this study, we define customers as citizens in Bandung and provider as Bandung government which provide public service. Involving citizens in co-creation process is like involving customers in product and service innovation in the service domain.

The governments need to clearly state what they and their citizens mean when they use the term 'increase choice' with the public services. Customers/citizens are a key aspect of public service delivery. Citizens can identify and define emerging and existing problems in the public services. They can conceptualize new solutions to well-defined problems there. They also can design or develop implementable solutions to well-defined problems there.

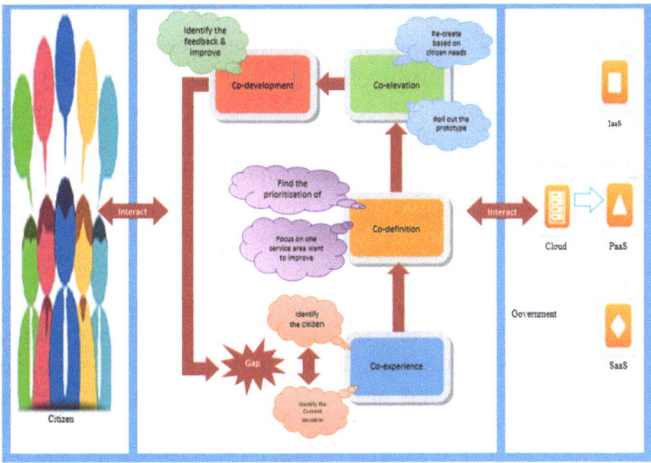

Fig. 10 Co-creation process between citizen and government

There are a lot of mechanisms can be employed by government agencies to facilitate an active role of citizens in innovating public services. These include mobile apps, e-petitions, open source databases, data analysis communities, open databases, participatory design workshops and dedicated online citizen communities. In this study, we propose co-creation process in Bandung. To be succeed in the co-creation process it is not enough to build and sustain a community, IT platform which provide the venue for citizen co-creation is needed.

By using cloud, government agencies have benefit to reduce the cost of IT ownership by consolidating their server through cloud. Moreover, the government programs become increasingly agile and responsive to change business condition, since by adopting cloud, the flexibility to deploy more current services with elastic capacity is allowed. Most of local agencies provide a variety of citizen services/ self-services. Open government is a classic example of informing and empowering the citizens through dashboards and scorecards about government and the flagship initiatives.

There is a great opportunity to learn more about how cloud computing is transforming the government in Bandung. The process of co-creation in detail is described in Fig. 10.

Based on Fig. 10, in the co-experience and co-definition process, the co-creation process by citizens needs to start with the identification of a need or problem faced by citizens. There are many urban problems in Indonesia, for example traffic jam, poor public service, waste disposal, natural disaster, crime and so on. These problems challenge the government, especially in urban area, to establish an independent state to become productive, livable and sustainable. Demographic, environmental, economic, political and sociocultural factors are forcing the world to become more efficient, dynamic, intelligent and self-sustaining or smart and sustainable.

In technical terms, there is a difference between the service level that each citizen should receive and the one that is currently receiving, referred to as the citizen service delivery gap. Probably, many citizen service delivery gaps identified need attention of the government. Local authorities or government need to prioritize them in order to focus on. By using cloud application, for example, it could be created for public service or social media channels for citizen to report the infrastructure project. Solutions to this kind of problems provide local authorities with higher flexibility when trying to tackle more wicked problems.

In the co-elevation process, it focuses on one service to improve. It is still possible to change the focus area; it allows a better interaction between the innovation team, government and citizens. After we focus on service to be improved, then a solution concept for the identified problem should be developed. This process should be centered on satisfying the needs that were previously identified. The actual creation of early prototype can be tested with potential users. Once the prototype is developed, it should be tested with actual citizens in a real world environment. In the case of public service, testing prototype is very useful in terms of the user experience and feedback.

In the co-development process, citizens who use the prototype should provide feedback about their experience using the service, and their feedback should be taken for the next iterations of the proposed solution. This feedback should then be used to improve the service. This iteration is critical, as it allows for improvements based on actual usage by citizens. To drive this proposed process, an innovation team needs to be established.

The value proposition offers are: firstly, secure system for internal governance data, including effective way of monitoring actual information for decision-making process for the mayor, growth curve of the city, efficient, convenient and paperless data administration. Secondly, a convenient and user-friendly system that is matched with the ICT level of government's human resource skill, appropriate for functional units communication and collaboration, and also met Bandung citizens' expectation of good online public service. Thirdly, citizens' engagement and participation with availability and secure feedback management.

The case study of this proposed model is the single ID card number (e-ID card) to improve the quality of public service in Bandung City by using cloud. This is related to the National's strategic programmes in population administration for the last three years, i.e., Population database update for residents across 497 regencies was completed in 2010; issuance of single identity number (SIN) for all residents in 330 regencies in 2010 and for all residents in 167 regencies in 2011; and the electronic ID program which is implemented in 2571 sub districts and 197 regencies in 2011 for 67 million eligible ID card holders.

Population administration issues are a very important step for Indonesia to curb the issuance of population and population database development. According to Law no. 23 of 2006, it is stated that the residents are only allowed to have one ID card. To be able to manage the issuance of ID cards and to realize a complete and accurate singular demographic database, it needs technological support such as cloud to ensure a high confidence in the identity of one's uniqueness and a high security identity data to keep away from counterfeiting and duplication.

In fact, Bandung residents may have multi-identity numbers. There are a lot of problem faced by citizen, i.e., multi data which is not integrated among the government agencies. The impacts are poor delivery service, long procedures and ineffectiveness since they need to search and input the residents' identity repetitively.

Therefore, government should meet the needs and expectation of the citizen by implementing the initiative single identity number which is known in Indonesia as Nomor Induk Kependudukan (SIN). It was a new method which is applied by government to build a new demography system and first formally established in the Senate Regulation Number VI/MPR/2002. It was later enacted in the Act Number 23, 2006, about population administration which was followed by the government regulation number 37, 2007, and the presidential regulations number 26, 2009, and number 35, 2010. The above regulations are the basis for implementing the single identity number and the electronic ID, also known as e-ID card programs.

The process of data recoding in the application of e-ID card in Bandung has been held since 2012. The number population of Bandung is about two millions and the ID card compulsory population who to data recorded target is 1.980,850 people. The implementing agency that handles nomenclature population and civil registry in Bandung city was population and listing of civil service under regulation of Bandung No. 13, 2007.

Population administration in Bandung is referring to National regulations number 7, 2009 (Law No.23, 2006 and Law No.37, 2007). The local government of Bandung has set 120 personnel in 30 districts who are temporary personnel during the data recording process. The implementation of e-ID card in Bandung is still lack of data recording devices, especially in the districts where they have a lot of ID card compulsory ($>60,000$). If there are no additional tools, 1.9 million persons ID card compulsory data recording in Bandung City will not be achieved as targeted. Based on the evaluation of implementation ID card, it shows Bandung citizens still not incomplete things to proceed e-ID card.

In this chapter, it emphasizes that citizen participation through co-creation process especially in urban as in Bandung is very important. Their willingness to cooperate with local government will get an added value. If people come in to do the recording photograph, signature, fingerprint and iris, then government programme will succeed. The important things as well as from local government side need to fulfil and innovate their public service delivery which based on the needs from citizen. Public cloud service as an enabler will work if both of parties participate in.

By providing the public cloud service from local government, both of parties will get benefit in value co-creation process. The local government should be more serious to build and to refine database system nationally as the forerunner of a single identity number (SIN). In order to deliver a good public service, government should develop an architecture design of open technology based (public cloud service) that can be implemented in a bottom up approach, value co-creation process which involve the citizen participation. The owner of e-ID card can be connected into a single National database, so that each citizen requires one ID card only (one map policy). The enabled cloud services which can be provided is e-voting, e-education, e-social welfare and e-health, which require all a single identity number and can access by different agencies of local government in Bandung.

Conclusion and Discussion

In this study, we propose how cloud can support Bandung based on value co-creation process, which involves the customers demand (society) to deliver improving citizen services. It is important to identify the gap between citizen needs and the current condition. Based on the current condition, the implementation of cloud computing in Indonesia, particularly Bandung, could be possible to be proposed, since cloud computing is an alternative technology for the improvement of ICT services for both government operations and public services.

Cloud technology can be proposed because it does not require large human resources. The level of services can be determined according to the current needs of the organization. Meanwhile, ICT personnel do not have to be provided in each SKPDs, and SKPD (functional unit) will be just the user of cloud services. Another thing is that cloud technology used will be the hybrid cloud.

One important thing before implementing a cloud technology is to identify the gap between current situation and the citizen needs. After finding the gap, it will be easy to focus on service that would be prioritized. For example in Bandung, it is important to deliver one-stop service. Internally, Bandung can gradually increase their quality of data administration, communication and monitoring within the government or even in every functional unit of Bandung government to bring a better performance and better decision-making process into reality. Externally, Bandung can always offer a better service to its citizens, including their potential business partners. In accordance with this, there will be a proposal of IT human capital training in the area of Bandung government and all its functional units in order that they will be able to learn on how to manage themselves in the provided system.

Limitation and Future work

The current study is an attempt to provide an updated outlook on the process of value co-creation by using cloud computing. As stated earlier, there is evident lack of knowledge mechanisms for value-co-creation in cloud-computing.

Therefore, this study describes value co-creation on cloud. However, cloud computing in this study is examined holistically, disregarding their various functions and areas of use. Therefore, it would be relevant to study (mostly for managerial purposes) how various cloud computing are used in collaboration. A promising avenue for future research would also be quantitative analysis.

Acknowledgement The author would like to express sincere gratitude and great appreciation to Prof. Suhono Harso Supangkat who have been contributing to the report undertaken by ITB Team. This material is supported by the final report of cloud community environment architecture in Bandung.

References

Andrei, T. (2009). Cloud computing challenges and related security issues; Anderson (2011), Springboard Research, White paper.

Armbrust, M., Fox, A., Griffith, R., Joseph, A. D., Katz, R. H., Konwinski, A., & Stoica, I. (2009). Above the clouds: A Berkeley view of cloud computing (Technical report no. UCB/EECS-2009-28), Berkeley: EECS Department, University of California.

Bappeda Kota Bandung. (2013). *Master plan Teknologi Informasi dan Komunikasi (ICT) Kota Bandung* 2013–2018

Bertalanffy, L. V. (1968). *General system theory: Foundatins, development, applications.* New York: G. Braziller.

Blazevic, V., & Lievens, A. (2008). Managing innovation through customer coproduced knowledge in electronic services: An exploratory study. *Journal of the Academy of Marketing Science, 36*, 138–151.

Brynjolfsson, E., Hofmann, P., & Jordan, J. (2010). Cloud computing and electricity: Beyond the utility model. *Communications of the ACM, 53*(5), 32–34.

Buyya R., Yeo, C. S., & Venugopal, S. (2008). *Market-oriented cloud computing: Vision, hype, and reality for delivering IT services as computing utilities.* Proceedings of the 10th IEEE international conference on high performance computing and communications, Dalian, China, Sept 25–27, 5–13.

Cambridge white paper. (2007). *Succeeding through service innovation developing a service perspective on economic growth and prosperity.* University of Cambridge Institute for Manufacturing (IfM) and International Business Machines Corporation (IBM), October 2007. ISBN: 978-1-902546-59-8.

Catteddu, D., & Hogben, G. (2009). *Cloud computing: Benefits, risks and recommendations for information security.* Greece: European Network and Information Security Agency (ENISA).

Christensen, C. M., & Rosenbloom, R. S. (1995). Explaining the attackers advantage: Technological paradigms, organizational dynamics, and the value network. *Research Policy, 24*, 233–257.

Dachyar, M., & Prasetya, M., (2012). Cloud computing implementation in Indonesia. *International Journal of Applied Science and Technology, 2*(3), 139–142; Centre for Promoting Ideas, USA.

Deighton, J., & Kornfeld, L. (2009). Interactivity's unanticipated consequences for marketers and marketing. *Journal of Interactive Marketing, 23*(1), 4–10.

Demirkan, H., & Delen, D. (2013). Leveraging the capabilities of service-oriented decision support systems: Putting analytics and big data in cloud. *Decision Support Systems, 55*(1), 412–421.

Edlund, Å. (2012). *Näringspolitiskt forum Rapport number 5*, Entreprenörskapsforum, 2012ISBN: 91-89301-44-7.

Galbrun, J., & Kijima, K. (2009). *Fostering innovation system of a firm with hierarchy theory: Narratives on emergent clinical solutions in healthcare.* Proceedings of the 52nd annual meeting of the ISSS, USA.

Grönroos, C. (2008). Service logic revisited: Who creates value? And who co-creates? *European Business Review, 20*(4), 298–314.

Haag, S., & Cummings, M. (2010). *Management information systems for the information age* (8th ed.). New York: McGraw-Hill/Irwin.

Haile, N., & Altzman, J. (2012). Value creation in IT service platforms through two-sided network effect. In K. Vanmechelen, J. Altmann, & O. F Rana (Eds.), *GECON 2012, LNCS, Vol. 7714*, (pp. 139–153) Heidelberg: Springer.

Hennig-Thurau, T., Malthouse, E. C., Friege, C., Gensler, S., Lobschat, L., Rangaswamy, A., & Skiera, B. (2010). The impact of new media on customer relationships. *Journal of Service Research, 13*(3), 311–330.

http://prnw.cbe.thejakartapost.com/news/2015/frost-sullivan-hybrid-it-and-cloud-enabled-technol ogies-will-define-the-new-normal.html

http://www.thejakartapost.com/news/2014/01/15/number-ri-internet-usersincreases-7119-mil lion-2013-apjii.html#sthash.dPY6ATYQ.dpuf

Jeffrey, K., & Neidecker-Lutz, B. (2009). *The future of cloud computing: Opportunities for European cloud computing beyond 2010.* 66, EU Cloud report.

Khajeh-Hosseini, A., Sommerville, I., & Sriram, I., (2010). *Research challenges for enterprise cloud computing.* ScoRR,a ans/1001.3257.

Li, Z., Chen, C., & Wang, K. (2011). Cloud computing for agent-based urban transportation systems. *IEEE Intelligent Systems, 26*, 73–79.

Lusch, Robert F., & Stephen L. Vargo. (2006). Service-dominant logic as a foundation for building a general theory. In Robert F. Lusch, & Stephen L. Vargo (Eds.), *The service-dominant logic of marketing: Dialog, debate, and directions* (pp. 406–420). Armonk: M.E. Sharpe.

Lusch, R., Vargo, S., & Tanniru, M. (2009). Service, value networks and learning. *Journal of the Academy of Marketing Science, 38*(1), 19–31.

Maglio, P. P., Srinivasan, S., Kreulen, J. T., & Spohrer, J. (2006). Service systems, service scientists, SSME, and innovation. *Communications of the ACM, 49*(7), 81–85.

Mangula, I., van de Weerd, I., & Brinkkemper, S. (2012). *Adoption of the cloud business model in Indonesia: Triggers, benefits, and challenges.* 14th international conference on information integration and web-based applications & services (hal. 54–63). New York: ACM.

Metcalfe, J. S., James, A., & Mina, A. (2005). Emergent innovation systems and the delivery of clinical services: The case of intra-ocular lenses. *Research Policy, 34*, 1283–1304.

Miller, M. (2008). *Cloud computing: Web-based applications that change the way you work and collaborate.* Indiana: Pearson Education, que publishers.

Normann, R., & Ramirez, R. (1993, July/August). From value chain to value constellation: Designing interactive strategy. *Harvard Business Review, 71*, 65–77.

Novani, S., & Kijima, K. (2012). Value co-creation by customer-to-customer communication: Social media and face-to-face for case of airline service selection. *Journal of Service Science and Management, 5*(1), 101–109.

Ostrom, A. L., Bitner, M. J., Brown, S. W., Burkhard, K. A., Goul, M., Smith-Daniels, V., Demirkan, H., & Rabinovich, E. (2010). Moving forward and making a difference: Research priorities for the science of service. *Journal of Service Research, 13*(1), 4–36.

Prahalad, C. K., & Ramaswamy, V. (2004). *The future of competition: Co creating unique value with customers.* Boston: Harvard Business School Pub.

Ristenpart, T., & Trommer, E et al. (2009). *Hey, You, get off of My cloud: Exploring information leakage in third-party compute clouds.* CCS'09. Chicago, Illinois, USA, ACM.

Sawhney, M., Verona, G., & Prandelli, E. (2005). Collaborating to create: The Internet as a platform for customer engagement in product innovation. *Journal of Interactive Marketing, 19*(4), 4–17.

Shen, Z., Wang, K., & Zhu, F, (2011). *Agent-based traffic simulation and traffic signal timing optimization with GPU.* Paper presented at the 14th international IEEE conference on Intelligent Transportation Systems (ITSC 11), Washington, DC.

Spohrer, J., & Maglio, P. P. (2009). Service science: Toward a smarter planet. In Karwowski & Salvendy (Ed.), *Service engineering*, New York: Wiley.

Vargo S. L., Lusch R. F. (2004, January). Evolving to a new dominant logic for marketing. *Journal of Marketing, 68*, 1–17.

Vargo, S. L., & Lusch, R. F. (2008). Service-dominant logic:continuing the evolution. *Journal of the Academy Marketing Science, 36*, 1.

Von Hippel, E. (1988). *The sources of innovation*. New York/Oxford: Oxford University Press.

Voona, S., & Venkantaratna, R. (2009). *Cloud computing for banks*. India: Infosys Technologies Ltd.

Willcocks, L., Venters, W., & Whitley, E. (2011). *Cloud and the future of business: From cost to innovation*. Accenture. (4 reports: Part 1: Promise. Part 2: Challenges. Part 3: Impact. Part 4: Innovation. Part 5: Management): Consulting reports.

Yazdani, B. (2012, March). *Building a cloud development culture* (pp. 36–39). Chief Learning Officer.

Zwass, V. (2010). Co-creation: Toward a taxonomy and an integrated research perspective. *International Journal of Electronic Commerce, 15*(1), 11–48, Taylor & Francis group, Abingdon.

Social Interaction in Travel Behaviour: Insights for Developing Effective Travel Demand Management for Indonesia

Yos Sunitiyoso

Abstract This chapter discusses research findings related to the effects of social interaction on travel behaviour. Insights gained from the findings are then analysed in relation to strategies and policies to manage travel demand, by reducing or redistributing the demand in space or in time for Indonesian context.

Introduction

Traffic congestion has become a major concern in Jakarta, Indonesia. Private car users and motorcycle users are dominating the roads. Public transport, which has a key role in transportation policies provided that it makes use of road space more efficient than a car, is not a preferable option due to lack of comfort, reliability and safety. If some car users could be persuaded to use public transport instead of the car, then the rest of the car users as well as the public transport users would benefit from improved levels of service as the traffic is less congested (assuming no further traffic is generated). However, car use provides the individual driver a number of immediate advantages which make the users unwilling to switch to public transport. Car is perceived to be a cheap form of transportation (Van Vugt et al. 1998), effective and efficient for multipurpose trips (Mackett 2003), and has a link to feelings of convenience and independence (Tertoolen et al. 1998).

Framing Transportation Problem as a Social Dilemma

The decision to commute by car or public transport not only has an impact on the individual commuter but also on other commuters. When more individuals commute by car on limited road spaces, people experience the negative consequences of

Y. Sunitiyoso (✉)

School of Business and Management, Institut Teknologi Bandung (ITB), Bandung, Indonesia

e-mail: yos.sunitiyoso@sbm-itb.ac.id

© Springer Japan 2016   65

K. Mangkusubroto et al. (eds.), *Systems Science for Complex Policy Making*,

Translational Systems Sciences 3, DOI 10.1007/978-4-431-55273-4_5

traffic congestion and environmental pollution. If more people decide to commute by public transport, this minimises the contributions to congestion and pollution. With regard to car sharing, there will be a reduction of the number of cars on the road if more people car share. If there is no change in travel demand, the number of cars in the road will be reduced to at least half of it if solo drivers can be persuaded to car share with other solo drivers. However, difficulties in practice may create barriers for people to car share. This particular type of interdependence with conflicting individual and collective interests can be framed as a social dilemma.

A social dilemma problem represents a conflict of interest between acting in public interest (cooperation) and acting in self-interest (defection). The social payoff to each individual for defecting behaviour is higher than the payoff for cooperative behaviour, regardless of what the other people do Dawes (1980). In travel mode choice context, this situation can be explained in a very simple situation of mode choice, which comprises car and bus as the only modes of transport. Social payoff in this context includes benefits such as less travel time, less congestion and more convenience. When some individuals choose to switch from car to bus, they will remove a number of cars from the road and therefore marginally reduce the traffic level and improve the journey for all users, as well as reduce environmental pollution. However, they gain a payoff less than individuals who do not switch to bus. Regardless of the decisions of others, car users always gain a better payoff. If all individuals are rational, then the best decision is to 'free-ride' on others by keeping up using the car. Second, if all individuals make a non-cooperative choice, then all of them will receive a lower payoff than the payoff if all individuals make a cooperative choice. This means that if all travellers use the bus, then they will receive a better payoff than if they all drive the car.

Solving the Social Dilemma

Transportation has an important role in maintaining the sustainability of our environment. The social dilemma in transportation is a key issue to be solved in order to reduce traffic and congestion. If traffic congestion can be reduced, then people can travel more efficiently (e.g. less travel time/cost, less environmental pollution) and the efficiency and viability of public transport services can be enhanced.

Van Vugt et al. 1998 categorised the ways for resolving the dilemma into two kinds of approaches. First, *structural approach* ('hard' measure), which includes interventions that alter the objective features of the decision situation by changing the incentive patterns associated with cooperation and non-cooperation, in other words, changing the structure of interdependence and effectively eliminating the dilemma. For example, by changing payoff structure (e.g. congestion charging), reward and punishment (e.g. incentives for public transport users, restriction on car parking) and situational change (e.g. residential or workplace relocation). Second, *psychological approach* ('soft' measure), which includes interventions aimed at influencing attitudes and beliefs that may guide people's cooperative and non-cooperative behaviours. This approach attempts to change the subjective interpretations of the situation, for example, by increasing individual's awareness of the environmental impacts of excessive car use (e.g. travel awareness campaign) and providing advice and information to encourage the use of alternative modes to the car (e.g. travel plan, individualised marketing). A demand management measure (or often called transportation demand management or TDM) is an application of a plan or policy aimed at changing or reducing demand for car use by encouraging the behavioural change of people's choices of travel. Some demand management measures have tried to address the social dilemma by incorporating structural approaches that offer material reward or punishment as well as psychological approaches that are 'softer' and more persuasive.

However, any intervention might not succeed since a strong inhibitor like habit can create resistance to change. Habit or habitual behaviour has been defined as choosing to perform behaviour without deliberation (Verplanken et al. 1997). Temporal structural changes using incentives (e.g. free public transport tickets for frequent drivers (Fujii and Kitamura 2003)) and changes of situational context (e.g. residential relocation (Stanbridge et al. 2004)) have the potential to break habits. Fujii et al. (2001) and Fujii and Gärling (2003) showed how cooperation can be triggered by freeway closure and a temporal structural change that functioned as a catalyst. A breakthrough to a new technological paradigm that provides new ways to satisfy mobility, such as internet and telecommuting, might also be a potential solution to break habits. Various demand management measures are aimed at breaking car-use habits. Some coercive measures would be more effective in breaking a habit, although not necessarily in yielding a new behaviour (Gärling et al. 2002). However, sustainable changes by individuals that can be integrated into their life patterns will be better achieved by persuading or encouraging rather than a coercive action, as Stradling et al. (2000) work revealed that in general motorists would rather be 'pulled' than 'pushed' from their cars.

Travel Behaviour is Dynamic

Travel behaviour has often been studied statically without considering that it is changing according to time, key events and other influential factors. This static assumption ignores that travel behaviour is the result of a complex and dynamic process involving interactions of many aspects and sequential adaptations over time (Fried et al. 1977). It also ignores that current travel behaviour results from past behaviour since past choices have also influences on the choices at present (Gärling et al. 1997). Individuals can and do change their behaviour over time. Fried et al. (1977) stated that the changes of behaviours may be due to internal processes (e.g. growth, life cycle) or external processes (e.g. technological changes, societal changes). Changes of environment require individuals to undertake dynamic decision-making processes which have three main characteristics: changing over time, availability of feedback and a need to make several interdependent decisions (Kerstholt and Raaijmakers 1997).

The understanding of how travel behaviour develops and changes over time is required in order to find the possibilities of creating positive changes and identify the potential effects of an intervention to change travellers' behaviour. The way of making decisions, as well as what and how influencing factors affect the process, needs to be investigated in order to get informed insights that can be utilised to predict the changes of behaviour. Although habits may be at work to inhibit changes, reasoned process has an important role when there is a need to make adaptation to changes in environment. For example, the introduction of a new mass transport mode (e.g. mass rapid transit/MRT) or a new car-use initiative (e.g. car sharing) creates a new option of commuting and can make some individual commuters change their choice of mode from car or bus to the new mode as it offers better features like reliability and punctuality. The behavioural change process is not a sudden process but a sequential and gradual process which requires adaptation and learning.

Travel choice is viewed as an adaptation to changes where people try out different choice options over time (Gärling et al. 2002). People need adaptation in order to solve the problem of how to learn and make a choice in uncertain and constrained environments. These threats create a challenge to travel behaviour research (Timmermans et al. 2003). Transportation system is highly dynamic, nonstationary and uncertain. Moreover, travellers' information is limited, imperfect and sometimes biased. Individuals do not know which choice is the best of his interest so that they decide to explore different choices in the beginning and become involved in more goal-oriented behaviour at later stage. However, this way is only effective if the environment is stationary. As most transportation environments are nonstationary, travellers may try different choices occasionally and later to develop an adaptive behaviour with the process of learning.

Social Interaction in Decision-Making

Structural interventions, which concentrate on changing personal material incentives (e.g. time, cost and comfort) associated with travel mode options, seem to be more effective since enforcements by the authority are also at force. Congestion charging has been proved to be successful in London, but it is still a controversial issue in other cities in the UK, and is still under government consideration to be implemented in Jakarta. In this situation, 'soft' measures, which do not incur additional economic costs on travellers, may offer an alternative option.

Provided that 'soft' measures are voluntary by nature and have no economic consequences (e.g. penalty from the authority) if travellers do not comply with the measures, 'soft' measures may become attractive for travellers. However, these characteristics have also drawbacks as it is difficult to ensure the sustainability of travellers' participation in the long term. Despite successful implementation of soft measures in pilot projects (Jones and Sloman 2003, Cairns et al. 2004), there remains lack of evidence on sustainability of impacts. Kitamura et al. (1999) argued that a demand management measure that relies on voluntary cooperation by individuals should be implemented only after it receives support from individuals that constitute the critical mass that is needed for prosocial behaviour to prevail. The critical mass depends on the conditional probability that an individual will cooperate, given cooperative behaviours of other individuals. Kitamura et al. hinted that understanding the characteristics of the interactive decisions made by individuals under the influence of other individuals is critically important for successful implementation of demand measures. Thus it gives an insight about the importance of social aspects in the implementation of a 'soft' measure.

The study of travel behaviour requires consideration that individuals' decision on whether or not to contribute to common interest depends not only on the past but also on their expectations to how their actions will affect those of others. This kind of relationship may exist and affect individuals' willingness to comply with a policy intervention.

Interdependence can be explained in a collective action (e.g. social dilemma) where there exists impossibility of exclusion, which means that no member of group engaged in collective action can be excluded from enjoying the benefits of the group's efforts (Huberman and Glance 1993). Personal choice or interest of an individual will not only affect herself but also other members of group. Messick (1985) defines interdependence in relation to preferences by stating the fact that individuals are not indifferent to the outcomes received by others.

Interdependence of choice may create a situation where travellers sometimes take into account and are concerned about choices by other travellers (Van Lange et al. 2000), which later develop as *expectations of others*. Some people will feel good in switching to public transport and doing their bit to fight congestion. Others will be influenced in their actions by how they will be judged by others (Lyons 2004).

The definition of interdependence may imply a relationship with the definition of expectations. Indeed they are really interrelated. However, interdependence refers to a situational context, where the choices of individuals are interrelated and affecting each other. They concern about the choices of others and make decision after considering others' decision, since their choices are interrelated. Symmetrically, others also concern and consider theirs. The 'quantification' or representation of the concern or consideration is formed as expectations.

Kitamura et al. (1999) raised the issue of the need to assess the effects of other users of the transport system who also respond to the demand management measures which highlight the importance of studying the effect of social interaction between travellers during the implementation of a measure.

It would be beneficial to study the social interactions of individuals inside a group or, in a wider scope, a society. Interaction and communication are likely to influence individuals' behaviour inside the group. Social interaction also contributes to the changes of environment. In his theory of social process, Douglas (1974) stated that social environment is constantly changing due to the contribution of individuals and groups engaged in social interactions.

Social interaction always exists whenever an individual is in an interdependence situation that involves other individuals. The scale of interactions depends on the size of group (or society). In a group, actions of a group member receive higher influence than those of in population, since inside a group there exists a feeling of belonging and responsibility as a group member. In population, those feelings may not strongly exist. An individual may expect that other individuals will take the action, so that she does not need to do anything.

Social interactions of members in a small group of individuals may be more beneficial and give more effects on changing behaviour. Within a small group, each individual has more responsible feelings about participating in a cooperative action according to the group interest, without ever thinking to 'free-ride' by being an opportunist since it will be easily seen by other group members. In relation to 'soft' measures such as the travel awareness campaigns, a more local and personalised campaign aimed at groups of people, for example, schools, companies or communities, may be more useful than a broad and national campaign aimed at whole population. Starting different projects and targeting different groups can give more effects on changing travel behaviour (Jensen 1999).

Olson (1965) stated that small groups are more likely to secure voluntary cooperation than are larger ones. By using an experiment of social dilemma with groups of people, they found two factors that can promote cooperation: repeated *social interactions* and *communication* among the participant. In mode choice context, more people might car share to work if they are aware of what others do and it can happen if there exist interaction and communication between them.

Ellison and Fudenberg (1995) studied the influence of communication structure in social interaction with what they called word-of-mouth communication. This way of communicating may allow efficient social learning process. Empirical studies by Ampt (2003) and Shaheen (2004) consider word-of-mouth communication as the way to diffuse the change of behaviour in voluntary car reduction programme in Australia and car-sharing programme in the USA, respectively. An empirical study in Japan also reveals the important role of social interactions through word-of-mouth communication. Taniguchi and Fujii (2007) in their study of promoting community bus service found that word-of-mouth advertising through recommendations to friends and family plays an important role in promoting bus use.

Fukuda and Morichi (2007), in their empirical analysis to measure and evaluate the effects of social interactions of bicycle parking (of railway users accessing stations using bicycles) in Tokyo, argued that variation in the collective behaviour across groups of users in different stations indicates the existence of social interactions among each group members (bicycle owners).

Based on these findings, in general, there are three levels of social interaction, which are being considered in this study. The first level of social interaction is due to an *interdependent* situation where none of individuals who engaged in a collective action can be excluded from enjoying the benefits/costs of their decisions (e.g. a social dilemma of public road users where the decision of each user affects not only herself but also the state of the system, hence affects other users). In other words, in an interdependent situation, impossibility of exclusion exists (Huberman and Glance 1993). The second level of social interaction may occur through a traveller's *observation* of other travellers' choices without involving processes of communication. The third level of social interaction is through *communication* between travellers regarding their travel choices. The last two levels of social interaction may, but not necessarily, happen between travellers at the same time as the first level of social interaction. Both the second and third levels of social interaction may be due to the fact that individuals are not indifferent to the outcomes received by others (Messick 1985) since travellers sometimes take into account and are concerned about choices by other travellers (Van Lange et al. 2000).

The Mechanism of Social Interaction

Travellers' choice making and behaviour can be considered as dynamic processes, since individuals can and do change their behaviours over time. The understanding of how travellers learn, develop and change behaviour over time is important in order to predict traffic congestion. A traveller's decision may be due to new information gained from their own experience, information and influence from the experience and behaviour of others through social networking, and official information from the authorities. These three types of information may be able to accelerate travellers' learning processes or shorten learning time required to make adaptation. They are also interrelated to each other and may be combined to make decisions. For example, official information from authorities may be spread to other travellers within a traveller' social network and combined with the traveller's own experience to make a travel choice.

Personal experience about the consequences of their decisions (e.g. the travel time that results from a particular choice of route) is essential within each traveller's learning process. The way these experiences might be used to inform travellers' future behaviour involves an explore/exploit trade-off, where the tendency of travellers to have favoured behaviours will need to be represented. *Social networking*, where travellers discuss their personal experiences of traffic conditions with each other, is an important aspect of traveller knowledge that has been little studied. Social learning is distinct from personal experience in that it allows travellers to build a wider (although possibly biased) view of the network without having to experience the conditions themselves and therefore enables a faster accumulation of knowledge and a consequent acceleration in travel pattern evolution. *Official information*, often in the format of online congestion maps, can provide accurate information on travel conditions across the whole network, but normally only at a high level of data aggregation and only for key routes (Fig. 1).

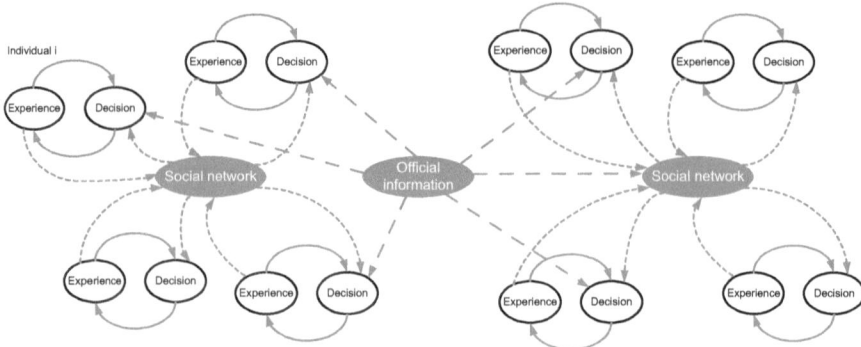

Fig. 1 Decision-making influenced by experience, social information and official information

Learning Individually and Socially

The concept of learning suggests that individuals learn from their past experience and acquire an adaptive decision-making process to cope with uncertain nature of environment (Timmermans et al. 2003). Combined with some social aspects considered in behavioural theory (e.g. beliefs and expectations, social norms, interdependency etc.), the concept of learning becomes wider in covering learning process from others' experiences or behaviours and in producing social learning concept. These learning processes are believed to be significant in decision-making processes of individuals in a dynamic choice situation.

There is a possible situation where learning may not occur because of habitual behaviour of travellers, where the decisions are made automatically according to situational cues and less deliberative. Learning is more likely to happen when there is a change of situational context (or goal/objective), when deliberation is prompted by information or when the situation is uncertain due to its nature or due to interdependence between people.

Learning can be based on individual experience (individual learning) and/or others' experiences or behaviours (social learning). Experience in individual learning can be used as feedback for the reinforcement of choice process (reinforcement learning), as feedback for evaluating decision rules (routine-based learning) or as feedback for evaluating cognitive model (associative learning).

In transport field, individual learning concept has been utilised in the context of mode choice, route choice and departure time choice. On the other hand, social learning has not been investigated intensively in the context of transport, although evidences from other disciplines (e.g. Pingle 1995; Pingle and Day 1996; Offerman and Sonnemans 1998; Smith and Bell 1994; Kameda and Nakanishi 2002, 2003) have shown that this kind of learning that is based on choices of others has been shown to be influential and important. Some exceptions in transport context are Ampt (2003), Shaheen (2004) and Sunitiyoso and Matsumoto (2009), although these research were not primarily aimed to study about social learning.

In social learning, decision makers may have the opportunity to observe the behaviours or preferences of others prior to making a choice; therefore, they can avoid decision costs associated with comparing alternatives. Also, the choice resulting from social learning may be high quality since it is learnt from other individuals with better performance in decision-making. Individuals can use several mechanisms in order to learn from others as suggested by Henrich (2004), including *conformist transmission* (imitate high-frequency behaviours), *payoff-biased transmission* (imitate people who are more successful), *self-similarity transmission* (imitate individuals with similarity at some traits) or *normative transmission* (follow the common behaviour in the group according to social norm). Among these mechanisms, conformist transmission and payoff-biased transmission are more widely studied (e.g. Smith and Bell 1994; Offerman and Sonnemans 1998; Henrich and Boyd 2001; and Kameda and Nakanishi 2002, 2003). The modeling of traveller's decision-making incorporating individual and social learning to cope with uncertainties and interventions are shown in Fig. 2.

In the literature, social learning is often confused with social imitation and the terms are often being used interchangeably; however, there is a slight difference between them. Bandura (1977) defines social imitation as a process in which individuals carry out the action that they observe from others without thinking much about the consequences (as found in the phrase 'when in doubt, follow others'), whereas social learning includes the consideration of the consequences of learned behaviour, instead of directly following others' behaviour. Individuals may learn from the mistakes made by others.

The difference between social learning and social imitation can be illustrated in the following example. In the beginning of study term, a new student followed the choice of travel mode of her housemates to go to university by bus since she did not have any idea and did not want to consider using other modes like cycling or walking. This part shows *social imitation*. After about couple weeks, she found that some of her classmates who lived nearby her house walk to university every day. Then she decides to walk as well since she thinks that it is healthier to walk and the distance is not as far as she thought and also she can save money. It shows *social learning*.

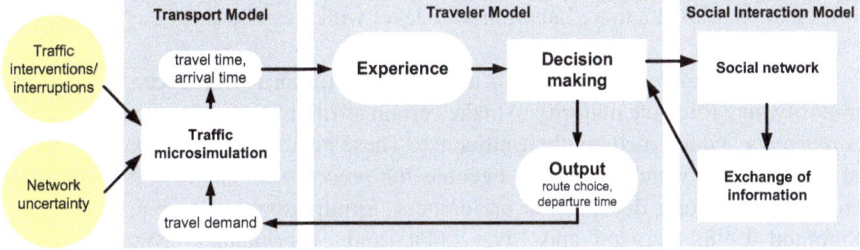

Fig. 2 Traveller's decision-making model

Majority and Minority Influences

According to Simon (1956) in his satisficing theory, it is stated that even if it appears that a decision is made by an individual independently of others, it often involves influence from family, friends or peers on reference group (e.g. schoolmates at school, colleagues at work) and also that different people in a group may have different weight (of influence), and this may change as the decision process evolves. This supports the argument about the importance of considering social aspects in studying travellers' behaviour.

Social influence has a strong link with social learning/imitation. The process of social learning and social influence may occur at the same time. However, there is a difference that can be drawn between social learning/imitation and social influence. In social learning, the change of judgments, opinions and attitudes of an individual is a result of active search for information by the individual, whereas in social influence, the change is a result of being exposed to those of other individuals (Van Avermaet 1996).

There are two types of social influence: majority (conformity) and minority (innovation). Majority influence can be shortly defined as the majority's efforts to produce conformity on the part of a minority, whereas minority influence can be defined as the minority's effort to convert the majority to its own way of thinking (Sampson 1991). Majority influence has a strong relation with a form of social learning transmission, the conformist transmission, which is a psychological propensity to preferentially imitate high-frequency behaviours or the most common (majority) behaviour.

In a situation where there exists minority influence, there are a few influential agents (independently or in group) which have more power to influence others whom they communicate with. The strength of their influence is derived from the reputation built from their consistency of choice to comply with the measure (Van Avermaet 1996; Sampson 1991, p. 155). An influential individual is not necessarily a traditional leader, but she can be a 'trusted person' with a respected reputation in the social club. Individuals are more willing to hear from someone who is trusted and respected as a consistent person.

In the minority setting, subjects have to deal with two groups – the minority and majority (Van Avermaet 1996). Under certain conditions a minority may not lead to public influence (because of normative pressures from the majority), but it could still be influential at a more latent, private level which can be shown as a change of preference value.

Consistency of decision/choice is the key in the minority influence. A consistent minority may force the majority to make certain attributions, such as *confidence* and *competence*, which mediate their influence. These attributions build the *reputation* of the minority which later may become *the power to influence* the majority to change its opinions, decisions or preferences. Simulation models (e.g. Nowak and Sigmund 1998; Gergaud and Livat 2004) and experimental economic games (e.g. Milinski et al. 2002; Haley and Fessler 2005) have indicated that reputation can sustain a high level of cooperation and prosocial behaviour. Fehr and Fischbacher (2003) also stated that experimental evidence strongly suggests that a considerable part of human cooperative behaviour is driven by concerns about reputation. A reputation for behaving cooperatively is a powerful mechanism for the enforcement of cooperation.

Latané and Wolf (1981), in the first principle of their theory of social impact (principle of social forces), argued that social impact is a multiplicative function of three factors: strength (e.g. power, expertise), immediacy (proximity in space and time) and size (number of the influence sources). The principle shares the same idea with Tanford and Penrod's (1984) theory of social influence model. However, Tanford and Penrod's theory is more formal, and it is based on certain features presumed to exist in any majority and minority influence situation, such as group size, number of influence sources, probabilities that a majority member will choose the minority position without any influence attempted by the minority and individual differences in members' susceptibility to persuasion. The minority influence may have smaller number of influence sources than majority, but it may still rely on strength (power to influence derived from reputation) and immediacy (closeness of relationships between individuals).

Finding and Its Implications

Sunitiyoso et al. (2009, 2010, 2011a, b, 2013) used various methods of study to understand the role of social information on travel behaviour. The study used behavioural survey to provide initial insights and to set parameter values for further stages of study. Then it utilized innovative methodology involving laboratory experiments to capture the role of the social interactions in the dynamics of human subjects' decision-making over time, which is extended with more individuals, longer repetitions and various settings in multiagent simulation experiments. Important findings of the studies are discussed with its policy implications.

People are Different: the Importance of Understanding Not Only Aggregate But Also Individual Behaviour

Sunitiyoso et al. (2011a, b, 2013) suggested that social interaction does produce differentiation of behaviour between different individuals or groups of individuals. By observing individual behaviour, it is found that a fraction of travellers is influenced in a positive direction (increasing contribution/cooperation) and the other fraction is influenced in negative direction (contribution/cooperation). Thus these two directions may cancel out each other resulting only in small changes (or no change at all) at aggregate level. The differences of learning process experienced by individuals can also be seen from the differences of the type of individual and social learning mechanisms used by the individuals. Thus it points out the importance of understanding individual behaviour (personally or in a group), not only aggregate (general) behaviour of individuals, since the changes of behaviour are often more observable at individual level.

It is found in the study that the effects of social interaction differ between individuals and groups of individuals. This may be due to the facts that individuals have bounded rationality, which are cognitive limitations and systematic errors (e.g. irrationality, biases) that constrain the individuals' judgement and decision-making. The limitations contribute to diversity and variability of the individuals' behaviour in response to the social information available to them. Other reasons may be due to differences in tastes or (random) experience. Some individuals and groups of individuals respond positively by increasing their contributions to a policy measure, whilst others respond negatively to the same policy measure by decreasing their contributions.

Travel behavioural change programmes often assume that people will change their behaviour if they receive information about why and how to reduce car use. However, this assumption is not always true. As suggested by Ampt (2003), understanding individuals is a key to successful travel behaviour change programmes. She stated that to *motivate* people, we need to understand that different people have different motivations, and it is a difficult task to motivate people to adopt some changes for reasons that are not important to all of them. Moreover, to *target* behaviour change that people will actually do and to *develop* behaviour changes that will appeal to people, we need to understand them. Some behaviour changes will appeal to people and others will not. Hence, to develop travel behaviour change tools that appeal to individuals, we need to understand that people are different.

Bamberg and Schmidt (1999) study shows that the same underlying motivational mechanisms caused different effects of the interventions for travellers to different objective settings. Generalising these results, one can conclude that the introduction of a measure in a different context (e.g. a different group) may yield different results. As also revealed in Sunitiyoso et al. (2011a, b, 2013), even the same policy interventions may affect people (as a group or individual) differently. The reasons may come from a conflict between the current state of an individual's behaviour and the changes promoted by a policy intervention. It may also be an effect of social interaction with other people which has influenced the individual's behaviour.

Sunitiyoso et al. (2013) revealed that the effects of social interaction through observation on others' behaviour are found no significant from the aggregated group behaviour point of view. However, there are changes of behaviour observed at the individual and group levels. This indicates the need to understand aggregated group behaviour as well as individual behaviour, since the effects of social inter-action may not be so visible at aggregate level but can be seen clearer at individual. There are turnovers between alternative choices of individuals in two directions (positive and negative) which can be 'symmetric', thus cancelling each other, or 'asymmetric', thus producing a change in either direction which may only look small and not significant at aggregate level. This phenomenon is called the 'churn' effect (Chatterjee 2001), which in travel behaviour context means that the changes may not be small changes in aggregate behaviour, but actually many individuals being persuaded to change their behaviour by changes to their environment.

The small changes at aggregate level may be due to the existence of individuals who maintain their 'status quo' of their (either high or low) contributions. Even the proportion of individuals who increase their contributions is larger than those who decrease their contributions; however, since there exists a significant proportion of individuals who are not affected, only a slight increase of contributions is observed at aggregate level.

In Indonesian context, the introduction of new public transport service, such as the bus service which runs on bus exclusive lanes (commonly named as 'bus way' in local word), was given high expectation initially to reduce traffic congestion in major cities, including Jakarta. The provincial government of Jakarta introduced this service in 2004. It has given some contributions in improving public transport services in Jakarta, although still too little to reduce the traffic congestion as the switch from car users and motorcycle users to bus users arguably is still minimum. The buses also often have to compete with other road users in their (supposedly) exclusive lanes due to opportunistic and impatient road users who are rather than switching to use the bus way, they chose to use its lanes and took the chance of getting a fine from the authority or even getting hit by the buses.

Customization of Policy Intervention

The study found that social interaction, learning and influence have different effects on different individuals. Changes may not be so visible in aggregated behaviour, but happen in individual behaviour. As an analogy in practice, a soft measure may have a less effect on an area with wide traffic levels, congestion or environmental impacts but it can succeed in changing target individuals who are using cars, therefore potentially resulting in, at least, benefits for individuals. In the long term, changes of behaviour are expected to diffuse to other individuals.

Policy interventions should be designed to influence behaviour in different ways for different groups of individuals. Thus it requires customization, different treatments/policies for different groups of people. Soft measures like personalised travel planning programme are targeting specific groups. For example, as discussed in Stopher (2005), 'Indimark' (individualised marketing) strategy is to segment population into several groups based on interests, willingness to consider and actual use of alternative modes to car. Then the treatments for each group are customised according to the characteristics of each group.

As indicated from the findings in this study, when provided with feedback about the outcome of their choices as well as social information about other individuals, some groups of individuals increase their contributions or maintain their high contributions (whilst others decrease or maintain low contributions). Thus a transport policy measure should be targeted to such groups. Goodwin (1999) indicated the need to target separately between people who are already stop driving and using public transport and people who are doing the opposite. For example, one of the focus in 'Travel Smart' programmes is on targeting the easiest car trips to shift, by the people most ready to do so. Therefore the main data required is to identify the likely switchers, monitoring being carried out in separate surveys, passenger counts and other measures as appropriate.

An initiative like Travel Smart is desperately needed in Jakarta to give people particularly businesses and employees a help to save time and money by helping to improve journeys to and from work to make Jakarta a better place to travel around. The initiative includes services such as journey planner for travellers, bus service improvement and bus vouchers, organising car sharing within workplace, facilitating walking and cycling routes and providing train services. Current efforts mainly focus on building the infrastructures, such as the MRT project, whilst other measures such as in Travel Smart are not yet available.

Identifying Target

In relation to the implementation of travel behavioural change programmes, policymakers have to be aware of both 'positive' and 'negative' effects of social interactions on different individuals or groups. Some people may be encouraged to participate in a travel behavioural change programme when they observe that other people are participating. Others may become discouraged and the rest may not react at all. Hence it is very important to identify, *a priori*, those groups of individuals where different reactions can be expected so the programmes can be customised, different treatments/policies for different groups of individuals. Also in real life, more 'positive' effects may be expected if the social interactions happen between close relatives, friends or colleagues who know each other rather than between anonymous individuals in the population.

As indicated in Sunitiyoso et al. (2013), some groups react positively to a specific intervention, other groups react negatively and the rest do not react at all. Bamberg and Schmidt (1999) stated the need to know how specific groups of individuals may react differently to the policy interventions such as heterogeneity could lead to a misinterpretation of the effects obtained. For example, car-sharing schemes can work when people are not all employed by the same organisation. However, it seems reasonable that schemes are more successful where their identity is one which people can more closely relate to. Liftshare's experience highlights that it is easier to find matches where schemes are for a specific community (Cairns et al. 2004). Ab Rahman (1993)'s review of car-sharing experience in the USA found a similar finding. Cairns et al. (2004) argued that travellers may adjust their behaviour in many different ways apart from the switch in mode of transport which is often the main focus of policy attention. For example, some travellers' response by changing the average distance of journeys, and in this case, it is not correct to calculate directly from mode switch to traffic impact.

Social interaction between individuals helps diffusing the information between them through various interaction domains ('social clubs'), such as neighbourhood, workplace, school, community and activity club, thus increasing the level of compliance to a demand management measure. In Indonesian context, community involvement has a crucial part for a successful implementation of a transport policy intervention (e.g. community-based car sharing).

Effect of Social Interactions through Social Information and Communication

Investigating the effects of information is also important, not only its existence. Sunitiyoso et al. (2013) found that, at aggregate level, social information about other individuals does not necessarily make people cooperate (choose to car share) more than when it is not provided. However, effects of the social information are observable at individual and group level. Tertoolen et al. (1998) revealed an interesting finding regarding the effects of information on travellers. When people were simply given information on their travel, and had no understanding of a way to change, they changed their attitudes ('travelling by car is not that bad after all') rather than their behaviour. Information about costs is not very likely to motivate travellers' action when the costs are low or not perceived as important by some people.

The study also indicated that the social information might have triggered competition rather than cooperation between travellers to gain more benefits than the others. When travellers' know what other travellers are doing, they react (according to their motives or expectations) by changing their choices, thus producing an instable situation. For example, when observing that most travellers are cooperating, travellers with a 'cooperative' motive or a 'bandwagon' expectation will also cooperate, whilst others with a 'selfish' motive or 'opportunistic' expectation will 'free-ride' on others. Then those who are cooperating may react to a selfish or opportunistic behaviour of the 'free-riders', thus producing instability.

Sunitiyoso et al. (2011b) found that participants were able to communicate with other travellers in the same group through communication using 'chat' service. However, the increase of cooperation (bus use) at aggregate level is not significant, although it significantly influences behaviour at group level. This may be due to an 'anonymous' communication between participants. Bartle and Avineri (2007) reviewed evidence that says face-to-face communication might be expected to have a greater influence on individuals' behaviour than, for example, anonymous email communication, or aggregated social information. Effects might also be different if participants already knew one another, in which case social norms and concerns about reputation might influence behaviour within the group. They identified some studies which have shown that cooperation in social dilemmas is improved when participants communicate with one another face-to-face. This implies that encouraging face-to-face communication, which is more likely to happen when people know each other such as between neighbours in neighbourhood, friends at school, colleagues at workplace and relatives, would increase cooperation.

The simulation experiments (Sunitiyoso et al. 2009, 2010, 2011a) provided behavioural insights regarding the diffusion of information using word-of-mouth communication as a form of social interactions, which enables social learning and social influence between travellers. The communication helps diffusing and maintaining sustainable participation in a travel behavioural change programme in a long period of time.

The insights support empirical findings of the role of *word-of-mouth* communication as a tool for diffusion. Ampt (2003) reported that in their voluntary behaviour change projects to date, it has been shown that messages delivered by any other way are reinforcing, but much less efficient. When the behaviour change has had a positive benefit to the individual, it is likely that they will tell others of the benefits (diffusion). Since they are more likely to practise diffusion in the company of trusted others, the message is more likely to lead to further change. Taniguchi and Fujii (2007) in their study of promoting community bus service found that word-of-mouth advertising through recommendations to friends and family plays an important role in promoting bus use.

Spreading information by 'word of mouth' may help solving a concern stated by Cairns et al. (2004) about the implementation of personalised travel planning. They stated that the effects may be short-lived if people quickly slide back into their old travel habits once the monitoring is over. There is some evidence suggesting that this may not usually be the case. In Perth, follow-up monitoring 2 years after the pilot individualised marketing project found the change in travel behaviour had been sustained (Ampt et al. 1998).

In the current era of social media, information and communication are seamless and without barrier so that 'word of mouth' can now switch to 'e-word of mouth' with the large number of Internet users especially social media users of around 40 millions in Indonesia. Traffic information site, such as the TMC Polda Metro Jaya (@TMCPoldaMetro), a Jakarta police department that has Twitter account, disseminates daily traffic information to around 3.75 million followers since 2009.

Customising the Information

It is often argued that giving information is a vital component of bringing about voluntary behaviour change. Whilst it is certainly important, it does not always motivate people to change. Information strategies develop out of the assumption that people will undertake the necessary actions once they know what they should do, how they should do it and why they should do it (De Young 1993). However, this is often not the case in reality. Providing information does not always produce the expected results. One of the reasons may be due to cognitive limitations of travellers in acquiring and processing the information, which result in diversity of behaviours between different travellers.

This highlights the importance of considering carefully the amount of information to be diffused since the information may produce some impacts which are not expected. Complete information is intended to assist travellers to make a better decision. However, unexpected results may be obtained. For example, Sunitiyoso et al. (2013) found that providing individuals with more (complete) information regarding other individuals' choices may have increased the level of uncertainty thus make them less 'cooperative' and less 'decisive'. Another reason may be due to travellers' cognitive limitations in processing large amount of information since it is also found that incomplete or limited social information produces a better result which is a higher level of cooperation.

In their study about marketing in the bus industry, Beale and Bonsall (2007) suggested the need to consider the different responses to come from different groups within a target audience. In order to avoid adverse impacts on people's bus use, the message content and target audience should be carefully considered as marketing may have a positive effect on some groups but a negative effect on other groups.

Role of Minority Influence in Social Interaction

Sunitiyoso et al. (2009) revealed that a rather small number of influential individuals (around 6 % of the population) are able to diffuse their choice to others and increase the level of compliance through various social interaction domains (e.g. neighbourhood, workplace, school). Also, a group that consists of influential agents is able to diffuse their compliance to other individuals of different groups. A 'social club' domain with a high opportunity of repeated interactions between its members, like course of study, has been reported to have an important role on the spread of compliance. Neighbourhood is a domain which has often been used in existing simulation models; however, it may have a smaller role than course of study since the interactions between neighbours, particularly in an urbanised city, are incidental and not as frequent as, for example, interactions within students in a course of study. These findings show that repeated interactions between individuals would generate high propensity for communicating which later give more opportunity for social learning and social influence (minority influence) to induce compliance in the population. Higher level of compliance (participation) with the measure is then produced with the support of 'minority' (influential) agents who are able to strongly influence other travellers' behaviour.

This study inferred that involving 'key people', which were represented as influential minority agents in the model, in diffusing compliance with the measure into population, would increase the level of participation. Ampt (2003) suggests a tool for changing travel behaviour is by involving key people early, not necessarily traditional leaders, but 'trusted others' in the community. The minority influence can be categorised as informational influence, which deals with the effect of knowledge, evidence or information in changing how one understands. With identification, the target accepts influence from the influential (minority) agent in order to preserve the strong bond of attraction that the target feels towards the agents. Then it is followed by internalisation which involves accepting influence from the agent because of its credibility and because it fits the target's existing beliefs and values. The minority influence may produce five possible responses on the behaviour of target individuals (Sampson 1991):

(a) Conversion: both publicly and privately accepts the position advocated by the agent
(b) Compliance: publicly accepts but privately not
(c) Anti-compliance: privately accepts but publicly not (common among major-ities who have been influenced by minorities but are either unaware or will not acknowledge it)
(d) Independence: both privately and publicly unaffected
(e) Negative conformity: changes initial position to even further from the agent's position

The first two responses are expected during the involvement of the 'minority' agents in promoting participation in a behavioural change programme. However, some people may also response the other way. Any of the other three responses may occur.

Identifying the influential individuals (the minority) is as important as involving them in promoting changes of behaviour using their influence. An influential individual is not necessarily a traditional leader, but she can be a 'trusted person' with a respected reputation in the social club as individuals are more willing to hear from someone who is trusted and respected as a consistent person. For example, a suggestion to car share by a consistent car sharer, who has been car sharing regularly in a considerable period of time, would have more influence than other individuals who have not done so.

In the simulation, the influential minority consists of agents (derived from respondents in the behavioural survey) who 'regularly car share and have been doing so for 6 months' and have high preference to car sharing. In practice, the minority could be some opinion leaders and credible sources who are drawn from the community and trained to do so. These influential people have a 'minority' influence which is strong in influencing change of behaviour of other people even though they are small in number. They can help people to overcome barriers to action and give ongoing support beyond their households. Jones and Sloman (2003) added the importance of involving key employees in the relevant organisations, so that they are aware of and supportive of the campaign in their dealings with members of the public.

Shaheen (2004) in her study on the dynamics of behavioural adaptation to a car-sharing programme suggested the need to study the effects of social interactions (e.g. of friends and family) on the diffusion process of a transportation innovation, since individuals do not make decisions in isolation but they frequently are moved to make these decisions in part as a result of social influence. Taniguchi and Fujii (2007) in their study of promoting community bus service found that word-of-mouth advertising through recommendations to friends and family plays an important role in promoting bus use. This interaction process might have begun a chain of bus use and recommendations. These studies highlight some indications that social interaction, social learning/imitation and social influence could be significant in encouraging a change of travellers' behaviour.

Social Learning Mechanisms: Confirmation and Conformity

An individual decides to make a change because of a variety of reasons. She may change at a point where the negative effects of an existing activity reach a certain level of intolerance and/or after observing someone else's behaviour. Kitamura et al. (1999) suggested that, in supportive circumstances, there are some possibilities such as people change their travel behaviours following others' behaviours.

Analysis of group behaviour in the experiments (Sunitiyoso et al. 2011a, b, 2013) indicates the existence of individual learning in the experiment. There is also an indication that giving participants access to social information about others' behaviours may influence their behaviours. It is supported by results of investigation on individual behaviour which shows that *confirmation* and *conformity* models are likely to exist whenever individuals have access to social information. It is also revealed that a stable/settled aggregate behaviour at system level does not necessarily correspond to stability of behaviour at individual level, which was endorsed by Goodwin (1998). However, when convergence exists at individual level as the results of a strong reliance on social learning, uncertainty in the system can be reduced as individual choices become more homogeneous.

These two types of social learning, *confirmation* and *conformity*, may exist in real life. An individual with *confirmation* mechanism of social learning imitates the behaviour of other individuals who have had similar behaviour in the past. For example, a car user may give more consideration on changing her travel mode to bus after observing her friends/colleagues who used to go by car change the travel mode to bus, rather than when observing those who are not car users. At the same time, an individual with *conformity* mechanism can be influenced by other individuals regardless of whether their behaviours in the past are similar or not with the individual. However, a critical mass of individuals who use bus is required since these individuals will only change their behaviour into the expected behaviour if the expected behaviour is the majority behaviour of other individuals. In practice, these two mechanisms may work together.

The study also provides indication that a strong reliance on *conformity* model of social learning may have a *positive effect* shown by increasing level of contribution. This is in line with the work of Sunitiyoso and Matsumoto (2009) regarding the role of *conformist transmission* in eliciting travellers' cooperation/participation. When travellers are not completely rational, they may have motivation to imitate the majority behaviour. This kind of social learning mechanism would give cooperation a chance to spread in the population. On the other hand, a strong reliance on *confirmation* model produces a *negative effect* shown by decreasing level of contribution. This indication could be true in real travellers' behaviour as social learning by *confirmation* may discourage people's cooperative behaviour whenever the observed individuals always choose 'uncooperative' choices.

In real situation, it is important to make sure that the positive effect of *confirmation* mechanism has a more prominent role than the negative effect (which occurred in the simulation experiment). In order to make sure that the positive effect has more prominent role than the negative effect, we need to be sure that the 'good' examples from influential people are spread through word of mouth to other people and give a strong influence to their behaviour. This again points out the importance of involving 'trusted others' that would spread the cooperative behaviour to their family, friends or colleagues. Their involvement would induce cooperation of a critical mass of people which will later encourage more people to cooperate through *conformity* mechanism (majority influence). In order to do so, we firstly need to identify these 'trusted others', which can be community leaders, 'diehard' cyclists or public transport annual ticket holders, and then get them involved in promoting the demand measures to other travellers. Then a mechanism, which is often used in commercial marketing like 'recommend a friend' or 'member gets member' scheme which can be voluntary or by offering incentives, would help the diffusion of the participation into wider population.

Implication on the Design of ATIS

The behavioural insights obtained in this study would have relation to the design of user communication features in advanced traveller information system (ATIS). In general, any types of ATIS have three main processes: collecting information, consolidating the information and then communicating the 'unified' information to various agencies and/or the public.

The study highlights the need for providing the 'right' amount of information to travellers as too much information may increase uncertainty. Customised information to specific group of people would be more effective both for the people and the transport system in general. The study also points out the need for studying the impact of other users of ATIS on a user's behaviour. A communication channel which enables the sharing of information between ATIS users regarding the users' experience on specific transport infrastructure or services may help other users to make a better decision. This is in line with Bartle and Avineri's (2007) argument that cooperative travel behaviour might be encouraged among users of ATIS, which should allow a greater number of people to make better informed travel decisions, which might in turn contribute to more efficient performance of the transport system overall. From a review of literature, they found that greater communication can lead to more cooperative/altruistic behaviour and enhance overall outcomes under certain circumstances such as when people wish to conform to a group norm which favours collective outcomes. However, there is also a possibility that in some cases, social interaction might lead to less cooperative travel behaviour, as indicated from the findings of negative effect of social interactions in the studies.

It is inferred from the study about the need to better understand the information acquisition process as more (complete) information does not necessarily improve travellers' ability in making decision but it may also unexpectedly create difficulties for them due to travellers' cognitive limitations in processing large amount of information. The simulation experiments provide behavioural insights on the long-term effects of a transport measure prior to its implementation in practice, which may also be applied in the study for investigating travellers' responses to ATIS over a long period of time.

A type of ATIS which now gained considerable subscriptions in Indonesia is the Google-owned maps and navigation app, Waze. By 2013, Waze had in estimate of over 750,000 active users and Indonesia is among the top 10 biggest Waze user communities in the world. Waze is available in the Indonesian language and there are also online user forums for local users.

Soft Measures in Combination with Hard Measures for Indonesia

In recent years, there has been growing interest in a range of transport policy initiatives which are widely described as 'soft' measures. Soft measures usually seek to give better information and opportunities which affect the free choices made by individuals, mostly by attractive, relatively uncontroversial and relatively cheap improvements. They include: workplace and school travel plans, personalised travel planning, travel awareness campaigns and public transport information and marketing, car clubs and car-sharing schemes, teleworking, teleconferencing and home shopping.

Despite the efforts to promote soft measures, only with them alone the traffic condition does not change much. In Indonesian context, 'soft' measures simply cannot be done without 'hard' measures through structural interventions. Those types of measures work as 'carrot and sticks' in an integrated transport planning.

The Carrots

The building of mass rapid transit (MRT) system, which is now undergoing construction targeting first-phase completion in 2018, is one of the 'carrots' that will improve public transport quality in Jakarta and attract car users to use public transport. With much higher carrying capacity and better quality of services than currently running TransJakarta bus services, the MRT system looks promising to reduce road traffic demands in the city. The provincial government of Jakarta is also planning to start its light rail transit (LRT) from 2016, whilst the monorail project is still pending.

The developments of public transport systems also need to be accompanied by other measures to promote alternative way of using car, as car use will still continue to be a dominant mode of travel. Car-sharing schemes shall also be further encouraged. They may include a range of different initiatives, including informal encouragement for arrangements for sharing trips which, to some extent, happen spontaneously anyway, between individuals at neighbourhood, workplace and even household level. Formal schemes, with elaborate arrangements for trip matching, often focused on commuting journey and organised linking with an ethos somewhat similar to hitch-hiking and often aimed at encouraging sharing for longer-distance leisure journeys. Some schemes are open to all and usually operate via Internet-based sites, whilst others involve initiatives confined to members of particular organisations and often combine websites with a more explicit management element. Car-sharing schemes may also offer incentives, such as high-occupancy vehicle (HOV) or 2+ lanes or reserved 2+ parking spaces for car sharers.

Cairns et al. (2004) revealed that in some schemes that offer few incentives, it is probable that publicity may encourage sharing although sharers may not necessarily be prompted to join an official scheme. Where there are incentives to join (other than to find a fellow sharer), it is probable that many existing informal sharers will join in order to capitalise on the benefits. It has also been found by Cairns et al. that it is important to have a critical mass of members registered on any car-sharing scheme in order to increase the probability of being able to make a match.

In Indonesia, car-sharing schemes are mostly based on community initiatives. A successful example is the 'Nebengers' car-sharing community (note: *nebenger* is a Jakarta' slang word for a person who likes to ask for a ride). It uses Twitter as the media for offering a share or asking for a share. This car-sharing effort has more than 10,000 active members and over 83,000 followers on their Twitter account. It is now not only operating in Jakarta, where it was started, but also other major cities in Indonesia. Despite community initiatives on car sharing, so far, there is no significant progress from the government on promoting car sharing as one of travel demand management measures.

Another soft measure that may also work in Indonesia is travel voucher. It is an employer-based transport initiative which asks each employee to use bus in every month. Countries such as the USA, Germany and France have adopted travel voucher schemes, such as Commuter Check and Travel Check (USA), Bonus

Malus (Germany) and the Carte Orange (France). The basic model is that the employer issues travel vouchers, to be used to pay for travel on public transport. These vouchers are tax-free, thus providing an incentive for the employers and employees to use them. Travel vouchers operate like luncheon vouchers, but for travel. Commuter Check (the trademark of one of the main travel voucher schemes operating in the USA) allows employees to use tax-free vouchers, given to them by their employers, on public transport. Employers can then claim the tax back from the government.

The travel voucher scheme might mostly encourage occasional new users of public transport (i.e. once or twice weekly bus trips which are shifted from cars) but might not trigger a large change in existing public transport users. Evidence from the USA (Root, 2001) shows that the introduction of travel vouchers usually has most effect on existing users of public transport, who mostly often increase their ridership when travel vouchers are available. This scheme has not been implemented in the Indonesia yet, but it can be done through workplace travel plans as long as incentives are provided by the government. As reported in Cairns et al. (2004), more employers could be persuaded to develop workplace travel plans if further tax incentives were offered. The incentives could be from business rate rebates, tax credits for travel plan revenue measures or enhanced capital allowances for infrastructure.

The Sticks

The well-known 'stick' in Jakarta roads is the 3-in-1 area in the major roads surrounding the so-called 'Golden Triangle' business districts which regulate that private vehicles with passengers of less than three persons are not allowed to pass the roads within the peak hours of 07:00–10:00 and 16:30–19:00. Its effectiveness to reduce the congestion is questionable due to its limited area and alternative ways to get around the area, including informal paid passengers. In future this will be replaced by an electronic road pricing (ERP) which will still take 1 or 2 years to be implemented. Higher parking charge has also been implemented in office and shopping districts to discourage private car users; however, due to lack of travel mode alternatives, car users are still willing to and can afford to pay the higher charges. Recently, the governor also announced the restriction of some major roads for motorcycle users in order to reduce the traffic due to its large number of users with more than 25 million operating in Jakarta.

These sticks will not work without the carrots. Improving transport systems means also changing the way people move. Sticks and carrots should work in harmony and complement each other. The road-building programme is also not the answer for anticipating traffic demand. In the opposite this may generate more traffic demand. Indonesian government should emphasise on building coherent transport networks with the road network, the rail network, bus services, walking and cycling and ports and airports, in order to meet the increasing demand for travel. Integration of transport policy with land use planning is also a key point that the government must be aware of as the traffic problem in Jakarta is largely caused by urbanisation. Concerns on environment must also be a main consideration by the government considering the amount of emission produced by transport and its potential impact to human health and its contribution to the climate change.

Acknowledgement The author would like to express sincere gratitude and great appreciation to Dr. Erel Avineri and Dr. Kiron Chatterjee who have been supervising and contributing to the studies undertaken by the author at the University of the West of England, Bristol, UK, in 2004–2007.

References

Ab Rahman, A. (1993). *Behavioral and institutional factors influencing car ownership and usage in Kuala Lumpur*, Ph.D dissertation, Texas A&M University Texas, 1993.

Ampt, E. (2003). *Voluntary household travel behaviour change: Theory and practice.* 10th international conference on travel behavior research. Lucerne, Switzerland.

Ampt, E., Buchanan, L., Chatfield, I., & Rooney, A. (1998). *Reducing the impact of the car: Creating the conditions for individual change.* PTRC European Transport Conference.

Bamberg, S., & Schmidt, P. (1999). Regulating transport: Behavioral changes in the field. *Journal of Consumer Policy, 22*, 479–509.

Bandura, A. (1977). *Social learning theory.* New Jersey: Prentice-Hall Inc.

Bartle, C., & Avineri, E. (2007). *Pro-social behaviour in transport social dilemmas: A review.* Paper presented at the workshop of frontiers in transportation: Social interactions. Amsterdam.

Beale, J. R., & Bonsall, P. W. (2007). Marketing in the bus industry: A psychological interpretation of some attitudinal and behavioural outcomes. *Transportation Research F, 10*, 271–287.

Cairns, S., Sloman, L., Newson, C., Anable, J., Kirkbride, A., & Goodwin, P. (2004). *Smarter choices: Changing the way we travel.* London: DfT.

Chatterjee, K. (2001). *Asymmetric churn – academic jargon or a serious issue for transport planning?* Available from http://www.tps.org.uk/files/Main/Library/2001/0001chatterjee.pdf.

Dawes, R. M. (1980). Social dilemmas. *Annual Review of Psychology, 31*, 169–193.

de Young, R. (1993). Changing behavior and making it stick: The conceptualization and management of conservation behavior. *Environment and Behavior, 25*, 485–505.

Douglas, J. D. (1974). *Understanding everyday life.* London: Routledge & Kegan Paul.

Ellison, G., & Fudenberg, D. (1995). Word-of-mouth communication and social learning. *Quarterly Journal of Economics, 110*, 93–125.

Fehr, E., & Fischbacher, U. (2003). The nature of human altruism. *Nature, 425*, 785–791.

Fried, M., Havens, J., & Thall, M. (1977). *Travel behaviour: A synthesized theory. Laboratory of psychosocial studies.* Chestnut Hill: Boston College.

Fujii, S., & Gärling, T. (2003). Development of script-based travel mode choice after forced change. *Transportation Research Part F, 6*, 117–124.

Fujii, S., & Kitamura, R. (2003). What does a one-month free bus ticket do to habitual drivers? An experimental analysis of habit and attitude change. *Transportation, 30*, 81–95.

Fujii, S., Gärling, T., & Kitamura, R. (2001). Changes in drivers' perceptions and use of public transport during a freeway closure. *Environment and Behavior, 33*, 796–808.

Fukuda, D., & Morichi, S. (2007). Incorporating aggregate behavior in an individual's discrete choice: An application to analyzing illegal bicycle parking behavior. *Transportation Research A, 41*, 313–325.

Gärling, T., Karlsson, N., Romanus, J., & Selart, M. (1997). Influences of the past on choices of the future. In R. Ranyard, W. R. Crozier, & O. Svenson (Eds.), *Decision making: Cognitive models and explanations*. London: Routledge.

Gärling, T., Eek, D., Loukopoulos, P., Fujii, S., Johansson-Stenman, O., Kitamura, R., Pendyala, R., & Vilhelmson, B. (2002). A conceptual analysis of the impact of travel demand management on private car use. *Transport Policy, 9*, 59–70.

Gergaud, O., & Livat, F. (2004). *Team versus individual reputations: A model of interaction and some empirical evidence*. 1st European Conference on Cognitive Economics (ECCE 1). Gif-sur-Yvette, France.

Goodwin, P. (1998). The end of equilibrium. In T. Gärling, T. Laitila, & K. Westin (Eds.), *Theoretical foundations of travel choice modelling*. Amsterdam/New York: Elsevier.

Goodwin, P. (1999). *Action or inertia? One year on from a new deal for transport*. Transcript of lecture given at Transport Planning Society Meeting at the Institution of Civil Engineers, 22 July 1999. London.

Haley, K. J., & Fessler, D. M. T. (2005). Nobody's watching? Subtle cues affect generosity in an anonymous economic game. *Evolution and Human Behavior, 26*, 245–256.

Henrich, J. (2004). Cultural group selection, coevolutionary processes and large-scale cooperation. *Journal of Economic Behavior and Organization, 53*, 3–35.

Henrich, J., & Boyd, R. (2001). Why people punish defectors: Weak conformist transmission can stabilize costly enforcement of norms in cooperative dilemmas. *Journal of Theoretical Biology, 208*, 79–89.

Huberman, B. A., & Glance, N. S. (1993). Diversity and collective action. In H. Haken & A. Mikhailov (Eds.), *Interdisciplinary approaches to nonlinear systems*. Berlin: Springer.

Jensen, M. (1999). Passion and heart in transport-a sociological analysis on transport behaviour. *Transport Policy, 6*, 19–33.

Jones, P., & Sloman, L. (2003). *Encouraging behavioural change through marketing and management: What can be achieved?* 10th international conference on travel behavior research. Lucerne, Switzerland.

Kameda, T., & Nakanishi, D. (2002). Cost/benefit analysis of social/cultural learning in a nonstationary uncertain environment: an evolutionary simulation and an experiment with human subjects. *Evolution and Human Behavior, 23*, 373–393.

Kameda, T., & Nakanishi, D. (2003). Does social/cultural learning increase human adaptability? Rogers' question revisited. *Evolution and Human Behavior, 24*, 242–260.

Kerstholt, J. H., & Raaijmakers, J. G. W. (1997). Decision making in dynamic task environments. In R. Ranyard, W. R. Crozier, & O. Svenson (Eds.), *Decision making: Cognitive models and explanations*. London: Routledge.

Kitamura, R., Nakayama, S., & Yamamoto, T. (1999). Self-reinforcing motorization: Can travel demand management take us out of the social trap? *Transport Policy, 6*, 135–145.

Latané, B., & Wolf, S. (1981). The social impact of majorities and minorities. *Psychological Review, 88*, 438–453.

Lyons, G. (2004). Transport and society. *Transport Reviews, 24*, 485–509.

Mackett, R. L. (2003). Why do people use cars for short trips? *Transportation, 30*, 329–349.

Messick, D. M. (1985). Social interdependence and decision making. In G. Wright (Ed.), *Behavioral decision making*. New York: Plenum Press.

Milinski, M., Semmann, D., & Krambeck, H. J. (2002). Reputation helps solve the 'tragedy of commons'. *Nature, 415*(6870), 424–426.

Nowak, M. A., & Sigmund, K. (1998). Evolution of indirect reciprocity by image scoring. *Nature, 393*, 573–577.

Offerman, T., & Sonnemans, J. (1998). Learning by experience and learning by imitating successful others. *Journal of Economic Behavior & Organization, 34*, 559–575.

Olson, M. (1965). *The logic of collective action: Public goods and the theory of groups* (2nd ed.). Cambridge: Harvard University Press.

Pingle, M. (1995). Imitation versus rationality: An experimental perspective on decision making. *The Journal of Socio-Economics, 24*, 281–315.

Pingle, M., & Day, R. H. (1996). Modes of economizing behavior: Experimental evidence. *Journal of Economic Behavior & Organization, 29*, 191–209.

Root, A. (2001). Can travel vouchers encourage more sustainable travel? *Transport Policy, 8*, 107–114.

Sampson, E. (1991). Innovation and the minority-influence model. In E. Sampson (Ed.), *Social worlds personal lives: An introduction to social psychology*. San Diego: Harcourt Brace Jovanovich, Inc.

Shaheen, S. (2004). *Dynamics in behavioral adaptation to a transportation innovation: A case study of Carlink – a smart carsharing system*. Institute of Transportation Studies, University of California, Davis.

Simon, H. A. (1956). Rational choice and the structure of the environment. *Psychological Review, 63*(2), 129.

Smith, J. M., & Bell, P. A. (1994). Conformity as a determinant of behavior in a resource dilemma. *The Journal of Social Psychology, 134*, 191–200.

Stanbridge, K., Lyons, G., & Farthing, S. (2004). *Travel behaviour change and residential relocation*. International conference on traffic and transport psychology 2004. Nottingham.

Stopher, P. (2005). Voluntary travel behavior change. In K. J. Button & D. A. Hensher (Eds.), *Handbook of transport strategy, policy and institutions* (pp. 561–578). Amsterdam/New York: Elsevier.

Stradling, S., Meadows, M., & Beatty, S. (2000). Helping drivers out of their cars: Integrating transport policy and social psychology for sustainable change. *Transport Policy, 7*, 207–215.

Sunitiyoso, Y., & Matsumoto, S. (2009). Modelling a social dilemma of mode choice based on commuters' expectations and social learning. *European Journal of Operational Research, 193*(3), 904–914. doi:10.1016/j.ejor.2007.10.058.

Sunitiyoso, Y., Avineri, E., & Chatterjee, K. (2009). The role of minority influence on the diffusion of compliance with a demand management measure. In: R. Kitamura and T. Yoshii (Eds.), *The expanding sphere of travel behaviour research*. Bingley: Emerald Group Publishing Ltd.

Sunitiyoso, Y., Avineri, E., & Chatterjee, K. (2010). Complexity and travel behaviour: A multi-agent simulation for investigating the influence of social aspects on travelers' compliance with a demand management measure. In E. Silva, N. Karadimitriou, & G. de Roo (Eds.), *A planner's encounter with complexity*. Farnham: Ashgate Publishers Ltd.

Sunitiyoso, Y., Avineri, E., & Chatterjee, K. (2011a). On the potential for recognising of social interaction and social learning in modelling travelers' change of behaviour under uncertainty. *Transportmetrica, 7*(1), 5–30.

Sunitiyoso, Y., Avineri, E., & Chatterjee, K. (2011b). The effect of social interactions on travel behaviour: An exploratory study using a laboratory experiment. *Transportation Research A, 45*, 332–344.

Sunitiyoso, Y., Avineri, E., & Chatterjee, K. (2013). Dynamic modelling of travelers' social interactions and social learning. *Journal of Transport Geography, 31*, 258–266.

Tanford, S., & Penrod, S. (1984). Social influence model: A formal integration of research on majority and minority influence processes. *Psychological Bulletin, 95*, 189–225.

Taniguchi, A., & Fujii, S. (2007). Promoting public transport using marketing techniques in mobility management and verifying their quantitative effects. *Transportation, 34*, 37–49.

Tertoolen, G., Kreveld, D. V., & Verstraten, B. (1998). Psychological resistance against attempts to reduce private car use. *Transportation Research A, 32*(3), 171–181.

Timmermans, H., Arentze, T., & Ettema, D. (2003). *Learning and adaptation behaviour: Empirical evidence and modelling issues*. Eindhoven: Behavioural Responses to ITS.

van Avermaet, E. (1996). Social influence in small groups. In M. Hewstone, W. Stroebe, & G. Stephenson (Eds.), *Introduction to social psychology: A European perspective* (2nd ed.). New York: Blackwell.

van Lange, P., van Vugt, M., & De Cremer, D. (2000). Choosing between personal comfort and the environment: Solutions to the transportation dilemma. In M. V. Vugt, M. Snyder, T. Tyler, & A. Biel (Eds.), *Cooperation in modern society*. London: Routledge.

Van Vugt, M., Van Lange, P. A. M., Meertens, R. M., & Joireman, J. A. (1998). How a structural solution to a real-world social dilemma failed: A field experiment on the first carpool lane in Europe. *Social Psychology Quarterly, 59*, 364–374.

Verplanken, B., Aarts, H., & Knippenberg, A. V. (1997). Habit, information, acquisition, and the process of making travel mode choice. *European Journal of Social Psychology, 27*, 539–560.

Unfolding the Problem of *Batik* Waste Pollution in *Jenes* River, Surakarta, using Critical System Heuristics and Drama-Theoretic Dilemma Analysis

Pri Hermawan and Ghita Yoshanti

Abstract Jenes River lies in the south of Surakarta City, passing through Laweyan, Serengan, and Pasar Kliwon. The condition of the water deteriorates day by day because of human activities including the process of *batik* making, textile factories, and neighborhood waste where most of the waste is directly thrown into this river. Before the water quality was degraded people could obtain many benefits from consuming the water and using it to support their daily life activities, but now because the river condition already below the quality standard, the water is no longer safe to consume and the river ecosystem is also damage. This chapter delineates this problem, stressing *batik* waste pollution as one of the causes for environmental degradation in *Jenes* River. Stakeholder analysis is explored with all possible claims from every stakeholder by using critical system heuristics (CSH). The claims are collected based on primary data from interview with several key persons, and secondary data from papers, journal articles, and newspapers. After displaying the problem constellation, conflict potency is explored using Drama-theoretic Dilemma Analysis (DtDA). DtDA is used to analyze dilemmas that arise in this problem. The DtDA analysis uses confrontation manager software. The results are several scenarios that can be implemented by the stakeholders to eliminate the dilemmas and propose solutions.

Introduction

Surakarta, which is renowned as a commercial city, is located in Central Java Province on the mountainous slopes of the Merapi and Lawu Mountains at a height of approximately 92 m above sea level; the area coverage is 44 km^2. Surakarta consists of approximately 500,000 people, most of whom are *batik* traders and workers. *Batik* is a traditional cloth-making process which use wax to make pattern.

P. Hermawan (✉) • G. Yoshanti
School of Business and Management, Institut Teknologi Bandung (ITB), Bandung, Indonesia
e-mail: prihermawan@sbm-itb.ac.id; ghita.yoshanti@sbm-itb.ac.id

© Springer Japan 2016
K. Mangkusubroto et al. (eds.), *Systems Science for Complex Policy Making*,
Translational Systems Sciences 3, DOI 10.1007/978-4-431-55273-4_6

93

The city produces reliable manufactured and commercial commodities, namely, hand-drawn *batik*, stamped *batik*, and printed *batik* (TREDA 2008).

The importance of *batik* as one of the commercial products of Surakarta has established several *batik* industrial clusters around the city. Porter gave definition to an industrial cluster as the geographic concentration of interconnected companies, specialized suppliers, service providers, firms in related industries, and associated institutions. Industrial clusters are widely considered to be an important means to promote regional innovation, entrepreneurship, and high-tech industries (Porter 2000).

In Indonesia, the agglomeration of small and medium-sized enterprise (SME) industrial clusters is observed in both rural and urban areas (mostly surrounding big cities). According to Weijlan, rural clusters in Indonesia have a seedbed function for the development of rural industries, demonstrating that clustering can improve access of rural producers to outside markets, through dense networks of traders (Weijland 1999). Klapwijk (1997) argued that clusters are important for the development of rural industries because productivity in clusters appears to be higher than that in dispersed enterprises. One of the main reasons is that clustering stimulates active involvement of traders and large enterprises (LEs) in agglomerations of SMEs (Weijland 1999). A more interesting finding is research from Sandee, which shows that enterprises in clusters are in a better position to adopt innovations in products as well as production processes than dispersed enterprises (Sandee 1996).

According to Tambunan (2000), most clusters in Indonesia were established naturally as traditional activities of local communities whose production of specific products has been proceeding for a long time (Tambunan 2000). Based on the comparative advantages of the products they make, at least with respect to the abundance of local raw materials and workers who have special skills in making such products, many of these clusters have a large potential to grow. Take, for example, the clusters of *batik* producers that have a long existence in various districts in Java (e.g., Yogyakarta, Pekalongan, Cirebon, Surakarta, Tasikmalaya). Various studies show the importance of clustering not only for the development of SMEs in the clusters but also for the development of villages or towns in Indonesia. The benefits of industrial clusters will help the stakeholders to overcome their constraints; to succeed in a more competitive market; to build an external and internal network; and to develop horizontal–vertical inter-firm cooperation-competition (*co-opetition*) (Bryant 2007).

The contribution to the high pollution in Jenes River comes from various sources, including the waste from textile factories, *batik* waste from small and medium enterprises, neighborhood waste, and hospital waste around the area through which the *Jenes* River passes. The damage caused by these sources makes the water harmful if consumed. The impact of the polluted river causes inconvenient conditions in the neighborhood. Before the water was degraded, people could use it for drinking, watering their plants, washing, even bathing, and various fish could live in the river, but now the water is darker, and has a bad smell. The nearby neighborhoods are also disturbed by the water and air pollution of the Jenes River. Several health impacts have also appeared; some of the neighborhood wells are already polluted, and if the people use the water, then their skin becomes

Table 1 Jenes River condition

Condition	BOD[a]	COD[b]	Copper (Cu)	Zinc (Zn)
Normal PP No. 82/2001	3 mg/l	25 mg/l	0.02 mg/l	0.05 mg/l
Jenes River	172.4 mg/l	383.7 mg/l	0.005 mg/l	0.019 mg/l

[a]*BOD* biological oxygen demand
[b]*COD* chemical oxygen demand
Source: Suara Merdeka (2003)

irritated and may have a rash. In the rainy season the water overflows and spreads around the neighborhood in the form of floods, so that conditions become very uncomfortable and the surroundings are not hygienic. The polluted water also can infiltrate to the land and contaminate the land quality. These conditions are based on reports in the local newspaper *Suara Merdeka* (24 February 2003) (Table 1). This table shows that the conditions are really problematic because the gap between the normal standard of water and the conditions in Jenes River is great.

This chapter mainly analyzes the impact of Jenes River pollution from *batik* production and the constellation of interests among the stakeholders. The claims among the stakeholders are explored using critical system heuristics (CSH). This method is part of system thinking methodology to unfold the simple coercive problems, mapping the stakeholders and their preferences or claims.

Research Objective and Research Questions

This research aims to unfold the problem around Jenes River pollution from various stakeholder points of view to understand the interrelationship, map out the problem, and explore the claims among the stakeholders. The elaboration will stress *batik* waste as one contributor to Jenes River pollution. This research explores the problem by using critical system heuristics (CSH); further conflict actor engagement is analyzed using drama-theoretic dilemma analysis (DtDA). This research aims to answer these questions:

1. How was the problem of *Jenes* River *batik* waste pollution elaborated from the multi-perspective stakeholder point of view (by using CSH)?
2. How can DtDA be used to analyze the dilemmas in this problem?

Literature Review

Critical System Heuristics

Several studies used CSH for structuring the problem, accommodating various perspectives, and fostering mutual understanding. Natural Resource-Use Appraisal (NRUA) in Botswana is one of the cases where CSH could help evaluate existing practices in natural resource use management for poverty alleviation (Reynolds 1998). Botswana is surrounded by a harsh environment of land covered by the Kalahari sands where it is difficult to practice commercially sustainable agriculture. Natural resource use involves difficulties in agriculture and wildlife utilization because surface water shortage results from low and variable rainfall patterns. The country has wealth from diamond mines as the wealth from the nonrenewable resources sector, but two-thirds of the population live in rural areas, variously engaged with livelihood activities based on renewable natural resources used. CSH here is used to support the intervention in two ways. First, within the framework of participatory rural appraisal (PRA) as an approach to structure and show the concerns of stakeholders, CSH is used to elicit and structure response from the stakeholders. Second, CSH is used to analyze the intervention and evaluate the use of PRA to assess the outcome and modify the participatory method in general. CSH evaluates PRA and outcome by revealing the limitations of NRUA from the claim of being inclusive and holistic; CSH as critical awareness also accommodates the marginal perspective and gives suggestions to improve participatory planning to be more responsible and professional (Reynolds 1998). Other cases also used CSH for healthcare planning; city and regional planning; energy and transportation planning (Ulrich and Reynolds 2010); enhancing prison service support (Flood and Jackson 1991); promoting an alternative lens for corporate responsibility; and informing international development initiatives (McIntyre-Mills 2004).

Using CSH in mapping out and evaluating the problem of *batik* waste pollution in *Jenes* River, Surakarta, is the first analysis in this paper. The current existing paper still elaborates the problem quantitatively or qualitatively for Surakarta or Indonesian cases, but the new approach presented in this chapter is the application of the CSH method in advancing problem identification and involving multiple stakeholders in the analysis, enriching understanding to find a better solution. The multiperspective solution is hoped to represent the various interests of all stakeholders and lead to holistic understanding.

Process of Drama-Theoritic Dilemma Analysis

DtDA takes the following steps (Hermawan et al. 2008b).

Identification of Crucial Stakeholders

Without oversimplifying the problem, the stakeholders are included in the conflict with each other in the current situation. Furthermore, to look at the problem in a holistic view, this paper tries to include those who are not directly involved in the confrontation but have interests in solving such problems (Hermawan et al. 2008a).

Identification of Position and Threat of Each Stakeholder

In this step the options (or strategies) are identified for each stakeholder to represent him or her in a "common reference framework." The beginning step before this step reveals the real situation of the conflict. In DtDA, how these available options and/or threats are adopted (or not adopted) by each stakeholder can give different views of dilemmas. Software such as Confrontation Manager is quite useful, especially when the problem involves a number of stakeholders and options (Bennett et al. 2001).

Identification of the General Structure of Dilemmas

Based on the positions and threats taken by each stakeholder in the current common reference framework, the paper identifies what dilemmas arise in such situations (Bryant 2003).

Scenario Generation and Analysis

Different scenarios are generated to investigate the basic underlying structure of the problem, and then to analyze the dilemmas and possible courses of action for the respective stakeholders to eliminate them (Hermawan and Kijima 2009).

System Thinking Methodology and Critical System Heuristics Method

System Thinking Methodology

In confronting changes in world phenomena, there are several system approaches in managing complex issues. One of them is used here to make a strong argument in analyzing the environmental impact caused by the *batik* industries in the Surakarta *Batik* Industrial Cluster. Critical system heuristics (CSH) was originally developed in the late 1970s by Werner Ulrich (1987). There is no clear answer to solve any certain complex phenomenon that happens in this world, and when we begin to initiate change unintentionally, consequences emerge. System thinking with all these approaches tries to simplify the process of our thinking and to manage the complex realities that are termed by system thinkers as a mess. A messy problem is mostly characterized as a problem with so many implications, so many people involved, with many interlocking aspects, a longer time scale, relatedness with interdependent factors, and extent of uncertainty, that it is difficult to figure out the answer or the end of the problems (Ackoff 1974). A messy or complicated problem should be addressed by system thinking to avoid a short-cut solution, because

Table 2 System thinking perspective based on situations

Perspective	Interrelatedness	Participants		
Based on problem situation		Unitary	Pluralist	Coercive
		'Hard' system based on machine metaphor	'Soft' systems based on organismic metaphor	'Critical' system based on prison metaphor
	Simple	Simple unitary:	Simple pluralist:	Simple coercive:
		System engineering	Strategic assumption surfacing and testing	Critical systems heuristics
	Complex	Complex unitary:	Complex pluralist:	Complex coercive:
		System dynamics and viable system model	Soft system methodology	Not available

Source: Jackson (2000)

conventional thinking is usually avoiding inevitable interconnectivity between variables, trapped in reductionism, and lies in single unquestioning perspective, which will lead to dogmatism (Table 2).

Critical System Heuristics (CSH) Method

CSH is used to understand more about interrelationships and interdependencies with multiple perspectives in a problem that needs to be structured. CSH was developed with the basic idea to support boundary critics to handle boundary judgment critically. Boundary judgment helps to determine the important, less important, and relevancies of empirical observations. CSH uses a set of 12 questions to make the everyday judgment explicit. The boundary questions try to figure out the situation around the system and cultivate a more holistic awareness of situations with regard to the complex issues. CSH tries to make sense the situation, to understand assumptions, and to appreciate the larger picture. CSH unfolds multiple perspectives delivered by different frames of situations and promotes mutual understanding of all the different points of view to handle the problem more constructively. CSH also tries to promote reflective practice by analyzing the situation and to change the perspectives by uncovering any undisclosed boundary judgment imposed on them.

CSH supports the system of intervention in two general ways. First, it can help to evaluate an intervention, and second, it can inform the methodologies used for intervention (Table 3).

Boundary critique in CSH is defined as a systematic (reflective and discursive) effort of handling boundary judgments critically (questioning one's claim before adopting it to others). The boundary critique of unfolding a boundary judgment is to achieve a 'whole system' view of a problem situation. The boundary judgment can

Table 3 Set of 12 questions in boundary categories of critical system heuristics (CSH)

	Boundary judgments informing a system of interest (s)			
Source of influence	Social roles (stakeholders)	Specific concerns (stakes)	Key problems (stakeholding issues)	
Source of motivation	*Beneficiary*	*Purpose*	*Measure of improvement*	*The involved*
	Who ought to be/is intended beneficiary of the system (S)?	What ought to be/is the purpose of S?	What ought to be/is S's measure of success?	
Sources of control	*Decision maker*	*Resources*	*Decision environment*	
	Who ought to be/is in control of the conditions of success of S?	What conditions of success ought to be/are under the control of S?	What condition of success ought to be/are outside the control of the decision maker?	
Source of knowledge	*Expert*	*Expertise*	*Guarantor*	
	Who ought to be/is providing relevant knowledge and skills for S?	What ought to be/are relevant new knowledge and skill for S?	What ought to be/are regarded as assurances of successful implementation?	
Source of legitimacy	*Witness*	*Emancipation*	*Worldview*	*The affected*
	Who ought to be/is representing the interest of those negatively affected by but not involved with S?	What ought to be/are the opportunities for the interests of those negatively affect to have expression and freedom from the worldview of S?	What space ought to be/is available for reconciling differing worldviews regarding S among those involved and affected?	

be questioned by doing individual reflection (boundary reflection) and dialogue with others (boundary discourse).

There are two kind of mapping in CSH: ideal mapping (ought) and descriptive mapping (is). The ideal mapping provides us with distance to what is really provided as point of reference for constructing and questioning descriptive mapping. Ideal mapping is to clarify the evaluator reference system, whereas descriptive mapping is to identify major boundary judgments built into the current situation. It starts with ideal mapping and identifies descriptive mapping. In descriptive mapping two steps can be done: identifying the stakeholder groups and specifying their role of concern.

Discussion

Stakeholder Mapping

Various stakeholders are involved in the problem of *batik* waste pollution in Jenes River. Some of the stakeholders were interviewed directly and the interview results used as the primary data in this research. Some of the other stakeholder points of view will be elaborated by using their preferences from secondary data of journals and newspapers. The interview process was conducted on 13 August–11 September 2014. The stakeholders and their stances were as follows.

1. *Batik* producers: In their opinion, because the government of Surakarta has issued a license for their production, and they still get the renewable license every year, it means that their production is safe for the environment. There are no major impacts to be considered and to stop their production. Some of the leading *batik* producers have developed their own wastewater treatment plant (WWTP). *Batik* Cluster Laweyan with the assistance of GTZ Germany developed a communal WWTP to be used by some of the *batik* producers around Laweyan. They do think about doing their best in preserving the environment, so they believe that the wastewater is not from their waste but comes from the waste of bigger factories in the upstream of the river.

2. Government (Central Java Governor, Surakarta Major, Parliaments, and Surakarta Environmental Agency): Realizing the importance of preserving the environment during production and the degraded condition of several rivers in Surakarta, the government sector amended the Central Java Regulation about wastewater quality standard no. 10, 2004 into UU no. 5, 2012. Then, the regulation about Industry UU no. 5 1984 was changed into UU no. 3 2014. In this regulation there are clauses about green industry that is environmentally friendly and sustainable requirements for conducting the production process. Green industry is an industry having has the green production process emphasis on efficiency and effectiveness in using raw materials sustainably, so it can put industrial development in line with environmental continuity and also benefit human life. Chapter 30 (1) said that the natural resources are processed and utilized efficiently, environmentally friendly, and sustainably. Chapter 30 (2) (a) - Chapter 30 (1) should be performed in the product designing process, production process planning, production process, waste optimization, and waste management. Chapter 79 (2) standard for green industry contains several steps: raw materials, production process, product, utilization management, and waste management. In Chapter 5, the government should define the wastewater quality standard for every activity and industry. They control the wastewater quality standard and ask for a report from the company once in a month.

3. Neighborhoods along Jenes River: The neighborhoods feel uncomfortable about the conditions in their surroundings; the people now need to buy drinking water because they cannot depend on their wells anymore but only use their well water

for cleaning. People who usually get their fish from the river cannot get these fish anymore because pollution makes it impossible for fish to survive in the river.

4. Local Non-Governmental Organization (LNGO): Help the neighborhoods to get a better environment for living by developing several programs to clean up the river and report any unusual condition about the river to the government to find a solution.

5. Academician: As the neutral stakeholder in this case, they want to propose the analysis to the other stakeholders, based on the academician point of view, supported by several theories and methods to target the problem and find a way out.

6. *Batik* consumer: In preferring to buy cheaper *batik* products, consumers are not really aware of the environment and the impact of their decision to buy cheaper *batik* products. Cheaper batik with chemical dyes uses materials and processes that are harmful toward the surrounding environments.

Critical System Heuristics Mapping (Ideal Mapping and Descriptive Mapping)

Dilemma Analysis

This section has applied drama-theoretic dilemma analysis (DtDA) to the Jenes River pollution problems (Tables 4 and 5).

Identification of Crucial Stakeholders

Based on the findings of the previous section, by using CSH stakeholder mapping, stakeholders that have critical dynamics in dilemma analysis are (1) government; (2) big factories; (3) small industries; (4) local NGOs; and (5) academicians.

Options available for each stakeholder are derived from the CSH descriptive mapping.

1. Government:
   • Do strict regulation

2. Big factory:
   • Build WWTP
   • Stop polluting the river

Table 4 Critical system heuristics boundary categories for ideal

	Boundary judgments informing a system of interest (S)			
Source of influence	Social roles (stakeholders)	Specific concerns (stakes)	Key problems (stakeholding issues)	
Source of motivation	*Beneficiary*	*Purpose*	*Measure of improvement*	*The involved*
	The neighborhood around Jenes River	To improve the condition of Jenes River so it can be beneficial for the environment and not causing disease	Cleaner Jenes River and free from pollution (good quality of Jenes River)	
Sources of control	*Decision maker*	*Resources*	*Decision environment*	
	The Government	The regulation maker and implementer	The environmental LNGO will criticize the way government makes and implements the regulation	
			The academician will report the observations based on the actual findings in the field	
Source of knowledge	*Expert*	*Expertise*	*Guarantor*	
	The academician including the natural and social scientists	Technical assistance, policy making mechanism assistance, social and environmental knowledge and skills	The suggestion and recommendation are quite applicable to apply and highlight the root causes of the problem, not only the symptoms	
Source of legitimacy	*Witness*	*Emancipation*	*Worldview*	*The affected*
	The collective citizen of Indonesia, the whole people of Surakarta, global presence in Surakarta and future generations	The better condition in Jenes River will bring health benefits for the surroundings and increase the quality of livelihood in Surakarta	Manage the conflicts among the stakeholder surrounding the Jenes River, for not pointing each other as the cause of pollution but working together for better Jenes River for the neighborhoods	

Source: Modified based on Jackson (2000)

3. Small industry:

- Build WWTP
- Stop polluting the river

Table 5 Critical system heuristics boundary categories for descriptive mapping

	Boundary judgments informing a system of interest (S)			
Source of influence	Social roles (stakeholders)	Specific concerns (stakes)	Key problems (stakeholding issues)	
Source of motivation	*Beneficiary*	*Purpose*	*Measure of improvement*	*The involved*
	The powerless neighborhood can only accept bad conditions	Help the neighborhoods to get a good-quality environment around Jenes River	Better health conditions and comfortable livelihood; now the situation is very uncomfortable, the quality of the river is also bad	
Sources of control	*Decision maker*	*Resources*	*Decision environment*	
	The Government, but the Government finds it difficult to implement the rules to powerful actors such as big companies (factory) in the upstream and other trade-offs; interest for increasing the production of *batik* for economic motives	The fairness in implementing the regulations, not hitting the powerless parties upon the powerful ones (ensuring fairness)	The action of LNGO and also media to check and balance of the government policy and implementation strategy	
Source of knowledge	*Expert*	*Expertise*	*Guarantor*	
	The natural and social scientists who want to find the solution for this problem (academicians)	Invention tools for tackling the problem and conflict resolution mechanism to stop the action of blaming one another and focus on the solution	Right innovation of WWTP with cheaper price and easily to implement; ensure fairness and not only forcing the powerless actors	
Source of legitimacy	*Witness*	*Emancipation*	*Worldview*	*The affected*
	The local and global present and the future generations	Better conditions will increase the livelihood and quality of life	Managing conflicts among the actors causing the pollution in Jenes River.	

Source: Modified from Jackson (2000)

4. Local nongovernmental organizations (NGO):

- Protest for a better environment

5. Academic:

- Find solutions

Identification of Position and Threat of Each Stakeholder

Position taken by each stakeholder and the threatened future are described as follows:

1. Government's position is:

    (a) Government should do strict regulation
    (b) Big factory should build WWTP; should stop polluting the river.
    (c) Small industry should build WWTP; should stop polluting the river.
    (d) Local NGO should not protest for a better environment.
    (e) Academics should find solutions.

2. Big factory's position is:

    (a) Big factory should build WWTP; should stop polluting the river.

3. Small industry's position is:

    (a) Small industry should not build WWTP; should stop polluting the river.

4. Local NGO position is:

    (a) Government should do strict regulation.
    (b) Big factory should build WWTP; should stop polluting the river.
    (c) Small industry should build WWTP; should stop polluting the river.
    (d) Local NGO should not protest for a better environment.

5. The Academic position is:

    (a) Big factory should build WWTP; should stop polluting the river.
    (b) Small industry should build WWTP; should stop polluting the river.
    (c) Academics should find solutions.

Identification of the General Structure of Dilemmas

The analysis is generated by using Confrontation Manager Software (demonstration version 1.3.1.13) (Fig. 1). The figure summarizes the general structure of dilemmas. In the figure, on the left side, the stakeholders are listed: government, big factory, small industry, local NGO, and academic. Each of them has two or more options. For example, small industry has option to "build WWTP" and "stop polluting the river."

The columns of the matrix represent positions that are offered by each stakeholder and the threatened future. The column under "Gov" shows the government's position, whereas that under "t" represents the threatened future. Other columns illustrate other stakeholder positions.

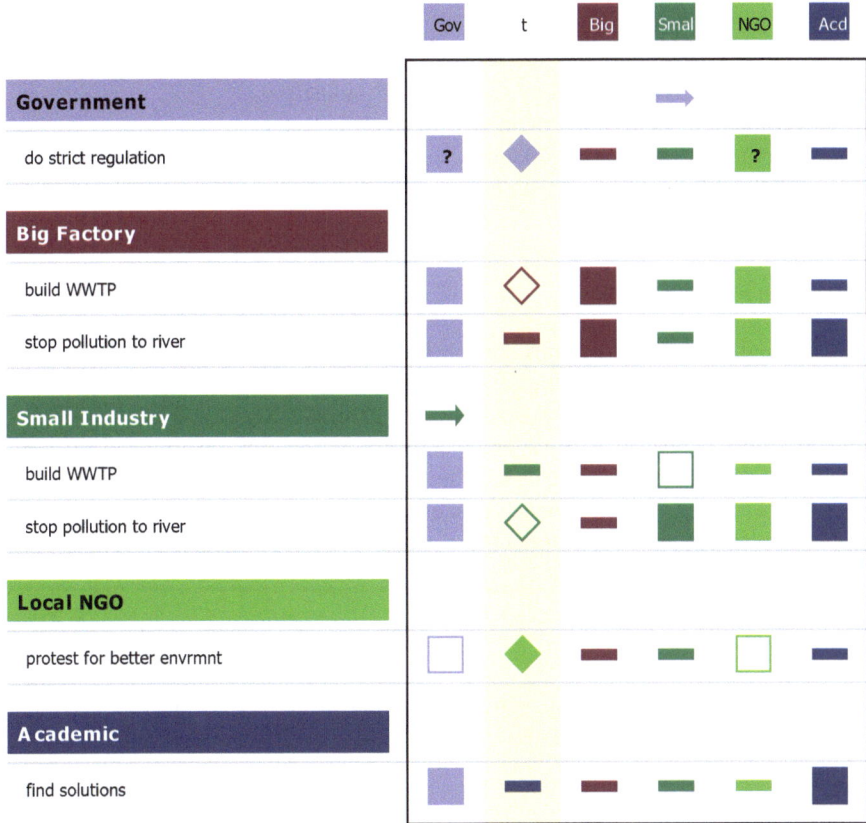

Fig. 1 Common reference frame for Jenes River pollution problem

In position or the threatened future columns, a filled square (or filled diamond, in case of threatened future) represents an option that is adopted by the respective stakeholder. Meanwhile, a white square (or white diamond) represents a position that is rejected by the respective stakeholder. A line means that the option is kept open, neither adopted or rejected by the stakeholders. For example, small industry's position is that "they would not invest in cleaner wastewater treatment and would stop pollute the river by untreated wastewater disposal, so that government should not enforce strict sanction on polluting."

An arrow, in this frame, represents preference between a position and the threatened future. The direction of the arrow shows which is considered more preferable by the respective stakeholder. For example, small industry prefers the threatened future to government's position, because the direction of the arrow in small industry's column goes to the threatened future.

A question mark identifies the doubt of one stakeholder's position respective option, at least by one other stakeholder. For example, government's threat to enforce strict sanction on polluting is doubted by the small industry and local NGO.

Now, we analyze some of the dilemmas arising in this situation. Based on the common reference frame, there are four dilemmas:

• Government's rejection dilemma with small industry
• Government's persuasion dilemma with small industry
• Government's cooperation dilemma with local NGO
• Local NGO trust dilemma with government

Government's Dilemma

Government has two dilemmas, rejection and persuasion dilemma with small industry. In the rejection dilemma, the government has a problem because their rejection of small industry's position is not credible. Regarding the persuasion dilemma, this happened because small industry rejects the government position. Small industry prefers the threatened future. Meanwhile, government cooperation dilemma with local NGO happened because the local NGO doubt that government would implement their commitment to perform strict regulation of pollution.

Local NGO Dilemma

Local NGO has one dilemma, which is trusting the government. This dilemma arises because local NGO doubt the government will actually implement its commitment.

Scenario Generation and Analysis

To eliminate the dilemmas arisen in this problem, some possible scenarios are proposed here.

Scenario for Government Dilemmas

There is common ground between government and small industry, in which they are agreed to stop polluting the river. For both, the small industry position is potentially better than the threatened future. Government can take conciliation or compromise as the course of action. The government's problem is small industry's insistence that small industry will not build a WWTP because it is very expensive.

Government can analyze small industry's underlying concerns. They can possibly send messages that suggest how to modify both positions to make them compatible. They can ask academics from the university to build an affordable WWTP process for small industry.

Regarding government's cooperation dilemma with respect to the local NGO, here the government problem is that the local NGO doubt that government would implement their commitments, if it is agreed. To eliminate this dilemma government must gain the trust of the local NGO. Government analyzes the local NGO assumptions. Why do local NGO believe that, if all parties agreed to government's position, they (government) might not do strict regulation? To answer this problem, government sends messages, by overcoming these assumptions, to do one or more of the following:

1. Show that the costs or difficulties government would incur in carrying out these commitments are less than local NGO suppose;
2. Show that the advantages they would gain from carrying them out are greater than local NGO suppose to gain;
3. Show that they must inevitably carry them out.

Scenario for Local NGO Dilemma

Local NGOs analyze government's concerns to see why after agreeing to the local NGO position, the government might not carry out strict regulation on polluting the river. Based on the discussion in previous section, if local NGOs could believe the government's commitment to do it, then that would eliminate the trust dilemma.

Conclusion

This case study gives an example of how to analyze a complex system in which many stakeholders were involved and participated. This paper proposed a stakeholder mapping process using CSH and then using a drama-theoretic general procedure to analyze dilemmas between various stakeholders, called drama-theoretic dilemma analysis (DtDA), to understand and, desirably, to solve the Jenes River pollution case in Indonesia.

As an implication of this research, by doing this systemic approach of stakeholder analysis to ensuring fairness among the stakeholders and resolving the dilemmas that arise in the interactions among them, we can identify barriers to collaboration. Moreover, as a general recommendation, we need commitment from all stakeholders and an effective multi-stakeholders' forum periodically, which should discuss the progress in implementing the results of the negotiation.

References

Ackoff, R. (1974). *Redesigning the future: A system approach to societal problems*. New York: Wiley.

Bennett, P., Howard, N., & Bryant, J. (2001). Drama theory and confrontation analysis. In J. Rosenhead & J. Mingers (Eds.), *Rational analysis for a problematic world revisited* (pp. 225–248). Chichester: Wiley.

Bryant, J. (2003). *The six dilemmas of collaboration: Inter-organisational relationships as drama.* Chichester: Wiley.

Bryant, J. (2007). Drama theory: Dispelling the myths. *Journal of the Operational Research Society, 58*(5), 602–613.

Flood, R. L., & Jackson, M. C. (1991). Critical system heuristics: Application of an emancipatory approach for police strategy towards carrying offensive weapons. *System Practice, 4*(4), 283–302.

Hermawan, P., & Kijima, K. (2009). Conflict analysis of Citarum River basin pollution in Indonesia: A drama-theoretic model. *Journal of Systems Science and Systems Engineering, 18*(1), 16–37.

Hermawan, P., Kobayashi, N., & Kijima, K. (2008a). Holistic formal analysis of dilemmas of negotiation. *Systems Research and Behavioral Science, 25*, 1–6.

Hermawan, P., Kobayashi, N., & Kijima, K. (2008b). Structure of dilemmas of negotiation: a drama theoretic model. *International Journal of Knowledge and Systems Science, 5*(1), 28–36.

Jackson, M. (2000). *System approaches to management.* New York: Kluwer Academics/Plenum.

Klapwijk, M. (1997). *Rural industry clusters in Central Java* (Institute research series No. 153). Amsterdam: VrijeUniversiteit.

McIntyre-Mills, J. (2004). *Critical system praxis for social and environmental justice: Participatory policy design and governance for a global age.* New York: Kluwer Academic/Plenum.

Porter, M. (2000). Clusters and the new economics of competition. *Harvard Business Review, 76* (6), 77–90.

Reynolds, M. (1998). Unfolding natural resource-use information systems: Fieldwork in Botswana. *Systemic Practice and Action Research, 11*(2), 127–152.

Sandee, H. (1996). *Small-scale and cottage industry clusters in Central Java: characteristics, research issues, and policy options.* Proceeding presented at the international seminar on small scale and micro enterprises in economic development anticipating globalization and free trade, SatyaWacana Christian University, Salatiga, November 4–5.

Suara Merdeka (2003) Pencemaran Melebihi Baku Mutu. Retrieved March 23, 2014, from http://www.suaramerdeka.com/harian/0302/24/slo3.htm.

Tambunan, T. T. H. (2000). *Development of small-scale industries during the new order government in Indonesia.* Aldershot: Ashgate Publishing.

TREDA (Trade Research and Development Agency). (2008). Indonesian Batik: A cultural beauty. In *Handbook of commodity profiletrade research and development agency.* Balitbangdag/PK/001/IX/2008. Retrieved on March 24, 2015, from http://www.kemendag.go.id/files/pdf/2012/12/08/batik-id0-1354950532.pdf.

Ulrich, W. (1987). Critical heuristics of social systems design. *European Journal of Operational Research, 31*(3), 276–283.

Ulrich, W., & Reynolds, M. (2010). Critical system heuristics. In *System approaches to managing change: A practical guide.* London: Springer.

Weijland, H. (1999). Microenterprise clusters in rural Indonesia: Industrial seedbed and policy target. *World Development, 27*(9), 1515–1530.